# THE ANCESTRESS HYPOTHESIS

THE RUTGERS SERIES
IN HUMAN EVOLUTION

Robert Trivers, *Founding Editor*

Lee Cronk, *Associate Editor*

Helen Fisher, *Advisory Editor*

Lionel Tiger, *Advisory Editor*

Other Titles in the Series:

Kingsley Browne, *Biology at Work: Rethinking Sexual Equality*

Kathryn Coe, *The Ancestress Hypothesis: Visual Art as Adaptation*

David Haig, *Genomic Imprinting and Kinship*

John F. Hoffecker, *Desolate Landscapes: Ice-Age Settlement in Eastern Europe*

John T. Manning, *Digit Ratio: A Pointer to Fertility, Behavior, and Health*

Paul H. Rubin, *Darwinian Politics: The Evolutionary Origin of Freedom*

# THE ANCESTRESS HYPOTHESIS

## *Visual Art as Adaptation*

Kathryn Coe

RUTGERS UNIVERSITY PRESS
New Brunswick, New Jersey, and London

Library of Congress Cataloging-in-Publication Data

Coe, Kathryn, 1942–
  The ancestress hypothesis : visual art as adaptation / Kathryn Coe.
      p.   cm.—(The Rutgers series in human evolution)
  Includes bibliographical references and index.
  ISBN 0-8135-3131-4 (cloth : alk. paper)—ISBN 0-8135-3132-2 (pbk. : alk. paper)
      1. Art and anthropology.   2. Art and society.   3. Art, Primitive.   4. Mother-
  hood—History.   5. Social evolution.   I. Title.   II. Series.

  N72.A56 C64 2002
  701'.03—dc21

                                                                    2002017875

British Cataloging-in-Publication information is available from the British Library.



Manufactured in the United States of America

*To my ancestresses:*
*My mother, Mary Ernest Jackson Coe*
*and my grandmothers,*
*Dora Jean Ellis Coe and Lizzie Loukate Wilson Jackson*

*To my codescendant*
*Anne Elizabeth Coe*
*and*
*To my two direct descendants and their spouses:*
*Blair Meredith and her husband, Cristoph Karl Schweiger*
*Trentham Parish and his wife, Victoria Mercedes Avella*
*And to their direct descendant and my more distant one:*
*Samuel Ernest*

# Contents

# Preface

I once spent the midmorning of a clear spring day interviewing Esteban Vicente of the New York School of Abstract Expressionists. As we sat in a garden next to a gallery filled with his paintings, I mentioned that the flowers blooming around us mirrored the beauty of the colors in his paintings. He glanced at the flowers and turned to me, saying, "Ah, my friend, only poets should write about art." I smiled at his cleverness in recognizing my poetic sensibility, nodded my head, said "of course," and did my best to write a poetic description of his work. This, however, is not a poetic book on art. This book is unabashedly theoretical and reductionistic and must be evaluated as a scientific study. It focuses, however, on topics often seen as frivolous or peripheral by the sciences: the visual arts, traditions, self-sacrificing mothers, and cooperation. While these topics may seem romantic, all of my assertions are cast in the form of falsifiable hypotheses. Visual art may be much more than what is discussed in these pages; however, the scientific method and modern Darwinian theory form the skeleton upon which this book is draped.

Although this book on visual art is reductionistic, condensed within that reductionism are the voices and behaviors of the indigenous potters, carvers, and weavers with whom I worked for more than thirty years in the southwestern United States; northern Mexico; small villages in Spain; Colombia's highlands; and Ecuador's paramos, coastal rainforest (Chachi and Tsachila), and Upper Amazon region (Canelos Quichua). Summarized here are decades worth of stories about creativity, mastery of technique, art history, the art market, and the politics of art told to me by contemporary painters, sculptors, photographers, and ceramists, both mainstream and regional. Without these colleagues this book would not have been written.

Five modest aims drove the writing of this book. First, I wanted to switch the theoretical focus of modern evolutionary theory from competitive, Coolidge-effect driven males to ancestral mothers and their competing strategies to see if such a focus would turn our assumptions of self-interest topsy-turvy. Second, I wanted to use visual art to test this ancestress hypothesis. Third, I wanted to concentrate on the persistence of behavior across generations; as George Williams noted, Darwin's theory is one to explain persistence more than it is change. Fourth, I wanted to show that the methods used by early ethologists—which focused on behavior, not mental processes—can help us increase our understanding of human behaviors, including cultural ones. Fifth, I wanted to build the argument that the effect of the inherited trait, not just the environment, influences the replication of the trait.

All these aims, as is probably often true, were products not only of my experiences but also my ancestry. Science gave me the ground rules. It also was the mechanism that allowed me to identify and juggle hundreds of distinct theoretical balls and bits of evidence as I attempted to elaborate and refine a collection of interconnected theories. Now, if this book attracts any attention, science hopefully will move this hypothesis forward, refining it through public debates grounded in empirical evidence.

Traditions attracted me, and I have devoted much of this book to their examination, because I lived many years among people who had, for various reasons, remained very traditional. Even though, it seemed to me, traditions often had significantly high costs, they were followed with care and protected with passion. When I returned to the United States and began to pursue my studies in social science, traditions became even more intriguing as they attracted so little theoretical attention. Traditions now seemed to be what every intelligent person since the Greeks had tried to escape or destroy. In his *Notebook,* Mark Twain bitterly described humans and their traditions: "We are," he wrote, "nothing but echoes . . . we are but a compost heap made up of the decayed heredities, moral and physical."

Unfortunately, as I lived among traditional people as complex, logical, and interested as anyone I knew, in empirical evidence, in identifying what worked and what did not, it was not possible to accept the explanation that they were prelogical, living perhaps in a world of spirits. Nor would it have been easy for me to accept that they behaved the way they did because powerful males coerced them. During the time that I had lived in small isolated villages or in isolated houses located along the banks of remote rivers, it was not clear who those males might be. There seemed to be no men, or women, who had the interest or the power to coerce others. There had to be, it seemed clear, another explanation for traditional behavior. This book is a culmination of my search for that answer.

I focused on mothers because my own mother and grandmothers were intriguing women who had me convinced that what men did, important as it may seem, was merely the means to an end. Fathers made it possible for a mother to weave her magic, as did my mother, by creating wonderful stories, writing and reciting poetry, growing flowers, piping roses on pink frosted cakes, designing jewelry, sewing costumes, playing anything on the piano that I wanted to hear, turning even a plain sandwich into a work of art, listening with sympathy to my petty sorrows, praising my occasional triumphs, and patiently guiding my often hesitant but occasionally exuberant steps into maturity.

My mother was largely a product of her mother, a sympathetic but no-

nonsense nurse, who had with sheer determination kept her poet husband alive for a decade after they were told that he was dying of tuberculosis. This grandmother was a social activist, a nurturer of her community who used to storm into the offices of elected officials demanding that they address (and waiting until they did) problems she saw in her community. Her house was a haven for children, filled as it was with hats and ribbons, costume jewelry, handmade lace, and silvery high-heel shoes. My sister Anne and I pulled felt hats with long, curling feathers down over our brows and draped ourselves in fox furs, laces, and satins, all so old that bits of them drifted in the air along with the dust motes that she told us were fairies.

My paternal grandmother was an intellectual, a college-trained teacher who had been raised on the Salt River Indian Reservation. Her social activism was expressed in the area of organizing literary clubs and helping build schools for American Indians. Her house was filled with books and anthropological treasures. There were Pima baskets woven with a portrayal of my great-grandmother and her dog and stern-faced idols from foreign lands sitting next to statues of the Virgin of Guadalupe with delicately carved fingers raised in prayer. There were inlaid boxes, bronzes, weavings, and baskets so large we hid in them.

My mother and grandmothers provided the evidence that I needed to know that motherhood was the world's most interesting job. How could I have ever doubted that art had its origin here, in the capable hands of women teaching and entertaining children and grandchildren? When I got to college, however, and read Virginia Woolf, I learned that women were seldom artists or, for that matter, creative beings. We did not have, I was told, the economic security that would make it possible for us to have a room of our own. Not only were we always controlled and confined, but we also were forced by the injustices of nature, at least until the pill, to be mothers. "Consider the facts," Woolf wrote, "First there are nine months before the baby is born, then there are three or four months spent in feeding the baby. After the baby is fed there are certainly five years spent in playing with the baby. You cannot, it seems, let children run about the streets. People who have seen them running wild in Russia say that the sight is not a pleasant one." If women were artists, she exclaimed, it "would necessitate the suppression of families altogether" (22). While Woolf seriously underestimated the amount of effort that raising a human child requires, the point is well taken that a room of one's own was a distant dream for most mothers throughout history. Even if it is true that the solitude such a room would offer may breed creativity, it also is true that humans made art long before any of our ancestors had rooms to themselves. Further, if archaeologists are correct, mothers, with infants at their breasts and children demanding their watchful attention, made a significant portion of the visual art made by early humans. Mothers and visual art were for me a natural connection when I was a child; now archaeologists have turned that natural connection into science.

My interest in visual art, while nurtured by my ancestresses, was further developed during the thirty-some years that I lived and worked with indigenous artists in Spain, Colombia, Ecuador, and the southwestern United States. It had became clear to me through various experiences that visual art was more important to them, that it played a more critical role in their social life, than I, or many others, had appreciated. Not only were crucial resources being used to produce arts that might soon be discarded, but when art objects were treated as if they were sacred, it was because of their association with the ancestors. Traditional visual art, and the ancestors who created it, seemed to be the warp and the weft, the threads that bound together so many elements of culture—kinship behavior, religion, politics, economics, moral systems, childcare—and that facilitated the transmission of knowledge from one generation to the next.

Although I now had some reason to believe that mothers played an important role in the evolution of modern humans, another issue remained: Is it true that all mothers, foolish and wise, indifferent and devoted, neglectful and attentive, good and bad, were all behaving selfishly? Was my childhood, which I remembered with fondness, actually one filled with unseen or unacknowledged conflicts of interest? What were my mother and grandmothers really up to? In particular, given that I should be taught to serve my own self-interests, why was a portion of each evening spent reading "tales of extraordinarily inspirational acts of courage and self-sacrifice" (Goodall 1999, 142–43). What were my mother and grandmothers trying to do to me by reading stories of Clara Barton; Florence Nightingale; the wife of Chief John Ross; Antigone; Judith; Mary, the mother of Christ, a self-sacrificing mother; the Spider Lady, who taught the Navajo to weave; the Clay Lady, who guided the hands of Pueblo potters; the Miguelitas, who fought beside their men in the Mexican Revolution; and Boadicea, who risked her life in battle to avenge the rape of her daughters.

The question, however, was deeper than this, as my parents were not alone in reading such stories to their children. Parents around the world over the centuries have read stories to their daughters about these female archetypes, dutiful daughter and self-sacrificing mother. Further, these women have inspired artists to produce powerful songs, stories, myths, poetry, and paintings about their lives. Why have the lives of these women been used to provide, generation after generation, role models for young girls and women? Why, despite the influence that these women have had on generations of women, young girls, and artists or poets, do they seem to have had such little influence on our theorists?

The answer to the first part of this question, it seemed to me, was that parents were encouraging their daughters to be generous and to care for the vulnerable; to be, in effect, a good mother and that this hypothetical good mother was very possibly the foundation of our moral systems and our social life. While defining the characteristics of a good mother may be something we are unable to do in our politically correct age, in myths and fairy tales, bad mothers are those who hold their own personal self-interests above those of their children, thereby neglecting, abusing, or abandoning their children. Good mothers were unselfish, in the sense that they put their children's interests before their own personal interests. The aim of the behavior of our ancestresses, who presumably were good mothers, was to do precisely what selfish genes would have them do; that is, behave in ways that promoted the persistence of their own genes in future generations.

While this selfish gene proposal logically must be correct, mothers did not get their genes into future generations by focusing on their own interests. A mother's interest is her child. There is a difference between the behavior of a mother who sacrifices her own interests, giving up her food or comfort so she can nurture, protect, and prepare her children for life, and that of a mother who takes resources away from her children so that she stays full and comfortable and is unburdened by the problems posed by children. There is a difference between a mother who takes the only blanket for herself and the mother who gives the blanket to a frail child. There is a difference between a mother who runs off to Hollywood to find herself and a mother who gives up such dreams. Yet, theoretically, all of these mothers are behaving selfishly. The use of the term *selfish* to describe all mothering behaviors made it difficult for us to distinguish between different maternal strategies and to identify any differential effects of the behaviors. The threads that we chose to use when we wove our definitions of the words *selfish* and *social* seemed to be too tangled to unravel.

Although I place great importance on mothers, I acknowledge that their importance is not obvious in our history books. However, female birds, to provide an analogy, are often drab compared with the males. This lack of decoration does not mean that females are unimportant. As a matter of fact, the decorated sex is trying, with some desperation, to attract the attention of the drab sex. This suggests to me that bright feathers do not determine ultimate value; the drab female is making the greater investment in offspring, and their (his and her) reproductive and dynastic success depends upon her doing so.

A sage once wrote that happy women, like happy countries, leave no history. As there is some truth to this, history often is about competition and conflict, why should we expect to find that the pages of history are filled with the activities of mothers who behaved in ways that promoted the interests of others? The fundamental question is this: Could self-sacrifice have been an important social strategy that was aimed at promoting the long-term success of our ancestors? In this book I attempt to answer the question of whether or not social behavior is characterized by selflessness and whether or not visual art was a mechanism used to encourage selfless or altruistic behavior.

To make this argument, this book is organized into five basic sections. In the first section (chapters 1 through 3), I examine the characteristics of traditional and nontraditional forms of visual art. In the second section (chapter 4), I use the characteristics of visual art implicit or explicit in this discussion to devise a working definition of visual art. In the third section (chapters 5 and 6), I outline the underpinnings of the ancestress hypothesis in terms of what factors promote reproductive and dynastic success. In the fourth section (chapters 7 and 8), the ancestress hypothesis is contrasted with Geoffrey Miller's sexual selection hypothesis. Predictions are made and tested against the evidence available. In the final section (chapters 9 and 10), I discuss modern Darwinian theory and attempt to reconcile it with the evidence found in traditional societies and, thus, the ancestress hypothesis.

A debt of gratitude is owed to Jane Newhall, who funded my research in Ecuador, and the department of anthropology at the University of Missouri-Columbia for granting me the time and space to write this book. I also wish to thank John Alcock, Matilda Essig, Mark Flinn, Christy Gallardo, Craig Palmer, and Clyde Wilson for reviewing chapters of this book, and my sister for raising important arguments. While their input was invaluable and their patience laudable, any errors, of course, are mine.

# THE ANCESTRESS HYPOTHESIS

To be simple is not always as easy as it seems.

—Ferdinand Hodler

It may be a wonder that any book on ancestors was written in the United States. We have no ancestors or, perhaps we have a great many, most of whom we have swept under the rug.

—Anonymous

# CHAPTER 1

# The Ancestress Hypothesis and Visual Art
## *An Overview*

What is art? Is it, as some have argued, man's noblest invention (Diamond 1992), or is it, as others have argued with equal seriousness, a mere by-product of other activities that are more important? Is art an activity that is spiritual and basically good and moral (John F. Kennedy referred to artists as engineers of the soul in 1963), or is it, as Socrates warily wrote, "the honeyed muse" (Plato 1977, 20) that feeds the passions and impairs reason and is likely to be, as Oscar Wilde once quipped, immoral? Is art, as some anthropologists insist, a necessary activity if we wish to promote the well-being of individuals and cooperation within societies; or is it, as Ralph Waldo Emerson wrote, a jealous mistress, robbing us of time and relationships? Is art a universal and ancient human activity, as some anthropologists argue, or is it (and the presumed aesthetic emotion), as some archaeologists hypothesize, found solely in Western societies?

These and other contradictory claims are the mystery that led me on a twenty-five-year quest to understand one of the arts, namely the visual arts. This book is a summary of that search, which, as luck would have it, was conducted precisely at a time when an understanding of visual art seemed at last to lie within our grasp. First, despite Marcel Duchamp's claim that art "has no biological source" (cited in Cabanne 1971, 100), modern Darwinian theory now provides us with a new vantage point from which we can view our species and its evolution and behavior, including art behaviors. Second, vast amounts of data have been accumulated—not only descriptions of contemporary forms of visual art but descriptions of visual art found in prehistoric, ethnographic, cross-cultural, and historic records. Thus we have both the theory and the data against which any hypotheses built upon that theory can be tested.

Within modern Darwinian theory, a number of proposals have been put forth to explain the arts. Hypotheses focus on art's neurological underpinnings (Aiken 1998); the role that art might play in devising scenarios to solve social problems (Alexander 1990); the importance of the arts in promoting cooperation in groups (Dissanayake 1992); and the evolutionary history of aesthetics or a sense of beauty (Thornhill 1998; Turner 1991). While these hypotheses are intriguing and productive, I wish to contrast my own with one that focuses on males, competition, and creativity. Geoffrey

Miller (2000), who argues that art is a strategy used to compete for mates, proposes such a hypothesis.

The evolution of art, Miller proposed, involves the evolution of a human tendency to make material objects into competitive and changing advertisements of fitness, of our appropriateness as mates. "As males tend to be the mate seekers, sexually mature males," he writes, "have produced almost all the publicly displayed art throughout human history" (Miller 2000, 275). "The human mind's most impressive abilities," Miller argued, "are like the peacock's tail; they are courtship tools, evolved to attract and entertain sexual partners" (4). Peacocks have brightly colored tails because the ancestresses of living peahens preferred males with gaudy tails." In a runaway type selection, humans began producing visual art and making it more elaborate. Elaborate art not only attracts attention, but it also has high costs. Only those individuals who are extremely fit are able to pay such high costs. Females, predisposed to select mates who could pay such high costs, produce sexy sons, sons more likely to be chosen as mates.

While it appears to be true that males (and even females) decorate themselves (and write sonnets, sing love songs, and show their etchings) when attempting to attract mates, the cross-cultural and archaeological characteristics of visual art suggest that much more than mate choice has been involved in the making and viewing of visual art. Humans, for tens of thousands of years, apparently have been decorating the bodies of infants, menopausal females, and the dead. More importantly, when visual art is viewed across the centuries and millennia, the majority of it, as I describe in this chapter and the next, has been traditional, not idiosyncratic. "Most writers," Radin argued, "have emphasized the conservation shown in the essential form of material objects" (1932, 53). Traditions imply replication or copying, not change or creativity. While perhaps technologically complex and attractive (meaning it attracts our attention), traditional visual art comes from ancestors and persists across generations. Nontraditional art, which explicitly ignores or rejects the traditions of one's ancestors, is characterized by individualism, creativity, change, and elaboration or even outrageousness. While Miller's argument may explain contemporary art, it cannot account for the traditional arts, and, as the majority of human visual art has been traditional, this is a serious defect.

While both Miller and I are evolutionary biologists, he places an emphasis on sexual behavior; I place it on parental care. Even though it must be true, as Miller argued, that human ancestors managed "to convince at least one sexual partner to have enough sex to produce offspring" (3), it is also true that if those offspring were to survive to reproductive age, a significant amount of resources had to be devoted to their care. While mating (just as survival) is necessary if humans are to successfully reproduce, it clearly is not a sufficient explanation.

The ancestress hypothesis, which is the name I give my argument, proposes that the linchpin of culture, and a driving force behind the evolution of our species, was the increasingly large investment made in offspring initially by mothers, and then allomothers, and then males. Humans are extreme K-strategists (investing large amounts in a small number of mates and offspring), as opposed to r-strategists (investing very small amounts in a large number of mates and offspring). Mating behaviors can be at the expense of the parenting behaviors that apparently have been so important to our species.

An ancestress strategy is not a female strategy; it is a maternal one. Mothers are not just females with children; they differ from nonmothers. "Pregnancy and motherhood," Hrdy wrote, "forever change a woman" (1999, 95). Motherhood, however, does not imply that one becomes an ancestress. An ancestress is a dynast; she is a woman who lived and reproduced and left a lineage of descendants influenced by her strategies. An ancestress strategy is long term, aimed not just at personal survival or procreation but also at using social strategies to promote the survival, reproduction, and social success of that offspring, its offspring, and their descendants.

While strategies that promote survival and reproduction can be, indeed are, self-interested, ancestress strategies are fundamentally social. A K-strategy is a social strategy. One crucial issue here is what is meant by the word *social* and whether or not social behavior is, like sexual behavior, at the expense of survival and whether or not it is, unlike sexual behavior, at the expense of increased reproduction. An r-strategy will out-reproduce a K-strategy in every generation.

If we assume that visual art not only influences behavior but is, as many argue, also an adaptation, then doesn't it make sense that it would be directed not only at courtship (which an be ignored), or mating (which may not lead to conception), or conception (which can fail to lead to a viable child), or even the birth of a child (who can die), but also at the survival and well-being of children and more distant descendants?

Adaptations, as traits that work over time, are apparently traits that originate, that promote reproduction, and that are transmitted to the next generation, thus promoting that generation's reproduction. Adaptations, in sum, promote not only the reproduction of individuals but also of their descendants. The proximate aim of visual art was to identify individuals who shared descent from a common ancestor and to encourage cooperative, unselfish behavior among all individuals so identified. Visual art's effect, the effect that promoted its persistence through time, was an environment in which large numbers of individuals who shared common ancestry, codescendants, identified themselves and cooperated as close kin (although many were *not* close kin) and thus were not threats to costly, vulnerable human offspring but were their protectors, providers, and teachers. The social restraint regularly encouraged by traditions would have been crucial. When traditions break down, so do the personal sacrifices they demand and the cooperative social relationships those traditions and sacrifices encourage. Art, increasingly, will become individualistic, competitive, and creative.

## Cross-Cultural Characteristics of Traditional Visual Art

In the archeological record we see remarkable persistence (meaning that one generation copies the prior generation) of material culture, including art. As Richard Alexander (1979) noted, culture can extend unchanged across multiple generations and far beyond an individual's lifetime. As M. G. Houston (1920, 2) has written specifically about art, "we are confronted with an extraordinary conservation or persistence of style, not only through the centuries, but through millenniums [*sic*]." Boas referred to this continuity in style as "fixed type" or "fixity" of design and form (1955, 144, 169). Despite Alexander's lament that he is "not optimistic about the usefulness of searches for unalterable or 'basic' human social behaviors as a method for solving our problems," the millennia that traditions have lasted suggest that humans often had fairly stable social strategies (1987, 9).

The Acheulian handax, although perhaps not an example of visual art, shows continuity of style (which apparently means replication of a particular form, motif, and technique) over approximately a million year period. Wood (1991) refers to the visual art produced by American Indians of the Plains over a 10,000-year period as showing "remarkable monotony of pattern" (33). Another example would be the Kakadu rock art of the Australian Aborigines, which shows continuity of a Rainbow Serpent design from 6,000 years ago to recent times (Mulvaney and Kamminga 1999, 359). Yet another example of continuity of style is found at Broadbeach Cemetery in Queensland, Australia. Here, the majority of mature males buried over a thousand-year period have the right upper central incisor removed ante mortem. According to the ethnographic record, dental ablation was a common part of male initiation; the particular tooth removed was determined by one's ancestry and served as an identifier of one's clan membership. As a significant number of the males buried at Broadbeach also share a dorsal defect of the sacral canal, we can assume they shared actual descent from an ancestor with those biological and cultural traits (Haglund 1976). Broadbeach, thus, was presumably a clan burial ground, a burial place for codescendants.

When we talk about persistence here, we are talking about traditions; that is, culture transmitted from ancestor to descendants, generally parent to child, over many generations. In this book, traditions refer *only* to behavior that is transmitted from one generation of kin to the next (see the glossary for definitions). In tribes and tribal subcategories (for example, clans, subsections, phratries, moieties), rights to particular techniques and design motifs are inherited, transmitted from one generation to the next in a lineage descending from a particular ancestor. Focusing on males, Morphy explained that subgroup affiliation was the way a man "obtains rights to produce his clan's paintings" (1991, 60–63). Meggitt described this in another Australian group: "The usual patrimoiety rules determine the manufacture of Gadjari headdresses and temporary ritual objects" (1965b, 228). Style, thus, refers to designs and techniques inherited from a common ancestor and shared by individuals who are codescendants. Whether intentionally or not, individuals are using visual art to communicate their ancestry and their relationship to others who share that ancestor.

When an individual accepts a tradition, and I use the term *accept* intentionally, as individuals have choices, that particular individual sacrifices or gives up freedom of expression at the ancestors' request. There is, as Boas pointed out, a "restriction of inventiveness" (1955, 156). When one accepts a tradition, one inherits the obligation to cooperate with one's elders in order to learn a design and the techniques necessary to produce it. One also inherits the obligation to teach these to the next generation. Further, one has to earn the right to use the art style one has inherited by showing appropriate social behavior. Morphy explained that a man must ask "his father's permission to do paintings that he has been taught" (1991, 61). If a male's behavior is seen as inappropriate, the ability to learn and to use a design is withheld: "Wamatun refused to teach his two eldest sons the clan's paintings unless they stopped drinking" (60–63).

Gombrich writes that the "great teachers of China thought of art as a means of reminding people of the great examples of virtue in the golden ages of the past" (1989, 104). Visual art often is associated with the religion and moral system, and it is this association that also is used to promote cooperation. One of the earliest Chinese book scrolls (fourth century C.E.) depicts the lives of virtuous ladies. Lega art consists primarily of human and animal figurines that are associated with proverbs about appropriate social behaviors (Biebuyck 1973, 45). Among the Australian Aborigines, Morphy explains, the "teaching of paintings is seen as part of the on-going process of initiations, and takes place in conjunction with the learning [from older male relatives] of songs and of some of the meaning of paintings" (1991, 60). Songs and stories describe how the ancestors in the paintings behaved and expect their descendants to behave. Ancestral heroes who lived in the dream time, Elkin writes, are models for correct social behavior. The actors in rituals, he writes,

> "dress up," adorning themselves with the paint and design peculiar to the rite, and wearing or carrying symbols which express some fact about, or incident in the life of the hero . . . At the conclusion of each act which usually lasts only five or ten minutes or so, the old men explain it and the decoration and symbols to any newly initiated men present or to any whose memories need refreshing. In this way, tribal history is handed down, and the patterns of life which the myths enshrine are instilled into the minds of the younger men present, for most do today what the great heroes did in the dream time. (1964, 156)

While the close linkage between kinship and art style helped promote persistence of visual art style, in ancient oriental urban societies, leaders and elders encouraged the replication of art traditions by arguing that "the traditional rules of art [were] as sacred and inviolable as the traditional religious creeds and forms of worship" (Hauser 1959, 31). When changes in style occurred, those changes often were adopted—as in the case of Christian art in the Middle Ages in Europe or the art of the Plains Indians—from metaphorical (or fictive) kin (Francastel 1967; Wood 1980). Metaphorical kin are not biological kin, but they use kinship terms to refer to one another and share kinship obligations. Further, the new elements of art complemented the ancestral style; they were not idiosyncratic; and they were justified by reference to ancestors (Fong 1984; Tonkinson 1978; Wadley 1997). As Tonkinson describes this, the Mardujara Aborigines claim that all new knowledge, including, presumably, knowledge of a new art style, comes to humans in dreams via spirit-beings who mediate between the living and their deceased ancestors (16).

Although one's ability to use traditions to promote one's own self-interest is clearly limited among the people described above, the costs associated with the traditional visual arts can be considerable. Decorating the body of dead ancestors can promote the spread of disease. Time is required not only to make an object but also to learn to make it properly. There are also energy costs, using energy that could have been used to perform other activities that have a more direct return. Further, getting the resources to make visual art often involves traveling long distances. The Maori sailed to remote areas of Otago and the west coast of South Island, either of which is quite a journey by sea, to obtain jade or nephrite in varying shades of green to make the hei-tiki, a small breast pendant or neck ornament of an ancestral female figure (Moore 1995, 176). Travel to get to sources of red ochre, at least in the case of the Australian Aborigines, could require crossing into enemy territory.

There also are other costs to one's health, particularly in the case of body

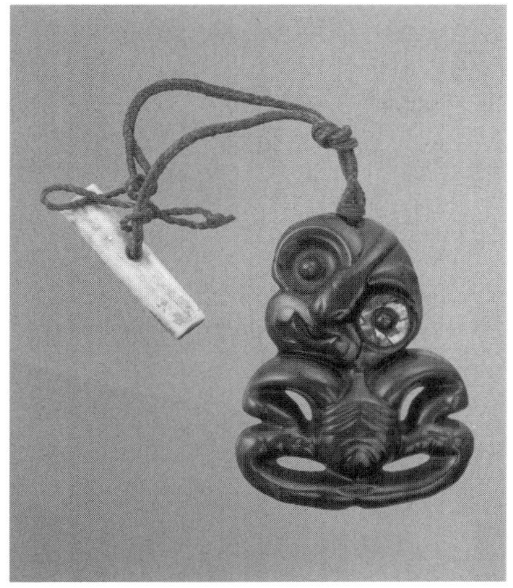

**Fig. 1.** Hei-tiki pendant (nephrite), Maori of New Zealand, circa eighteenth century. Courtesy of the George Ortiz Collection, Switzerland.

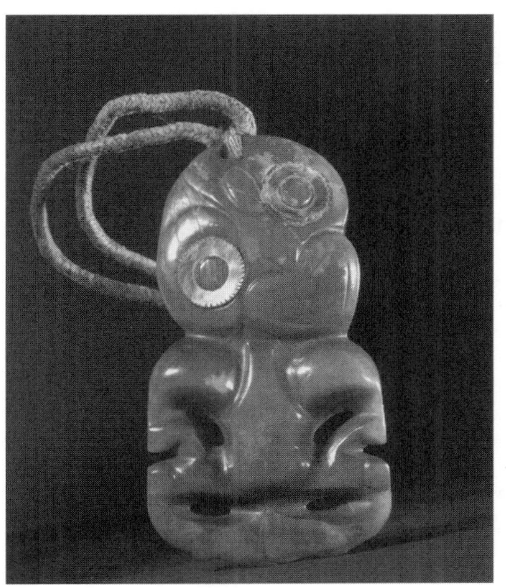

**Fig. 2.** Jade hei-tiki or breast pendant from Maori of New Zealand, circa eighteenth century. Prized as personal ornaments, some hei-tiki also became sacred clan heirlooms, accumulating the mana of each owner. Courtesy of the British Museum, London, and the Werner Forman Archive/Art Resource, New York.

decoration, which often is referred to as one of the earliest forms of visual art (Weltfish 1953). Tattoos and scarification, for example, are painful and can lead to infection. Dental decoration (for example, filing, ablation, inlay) can result in alveolar abscesses and bone infections, are more easily subject to wear, and can cause problems with speech and mastication (Linne 1940; Romero 1970). Intentional cranial deformation can cause exotoses in the auditory canal (Hrdlicka 1940), decrease cranial volume (6 percent in some Peruvian skulls [MacCurdy 1923]), change the shape of the

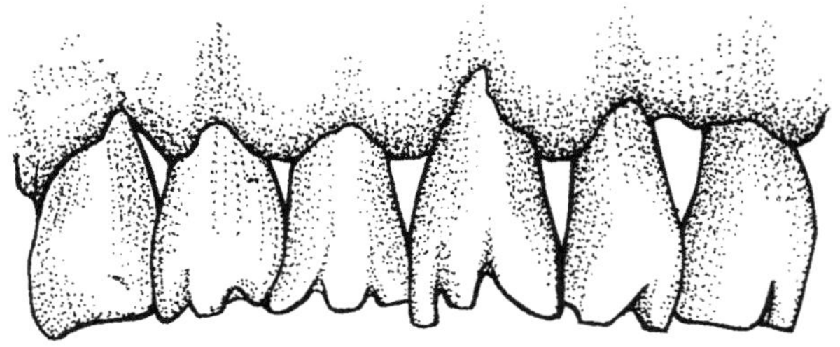

**Fig. 3.** Teeth filed for decorative purposes. Drawing by Matilda Essig. Adapted from Vera Tiesler Blos, *Decoraciones dentales entre los antiguos mayas* (Mexico, D.F.: Ediciones Euroamericanas, Instituto Nacional de Antropología e Historia, Cordoba, 2001).

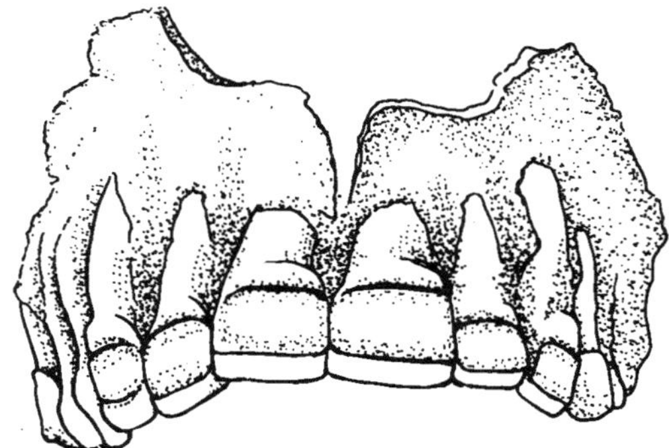

**Fig. 4.** Teeth filed for gold lamination. Illustration by Matilda Essig. Adapted from J. Romero Molina, "Nuevos datos sobre la mutilación dentaria en Mesoamérica," *Anales de Anthropología* 23 (1986): 239.

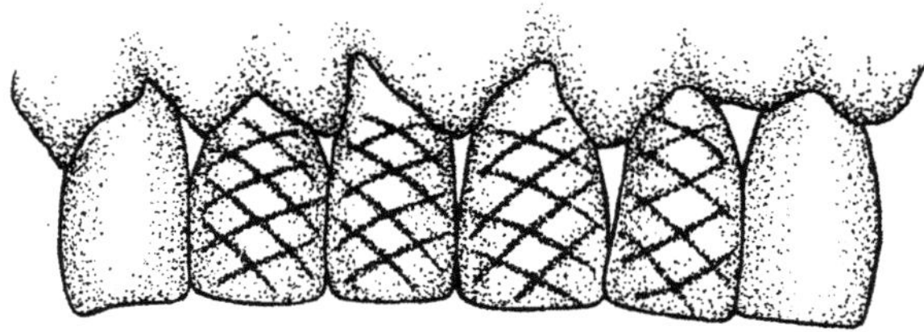

**Fig. 5.** Incised teeth. Illustration by Matilda Essig. Adapted from J. Romero Molina, "Nuevos datos sobre la mutilación dentaria en Mesoamérica," *Anales de Anthropología* 23 (1986): 239.

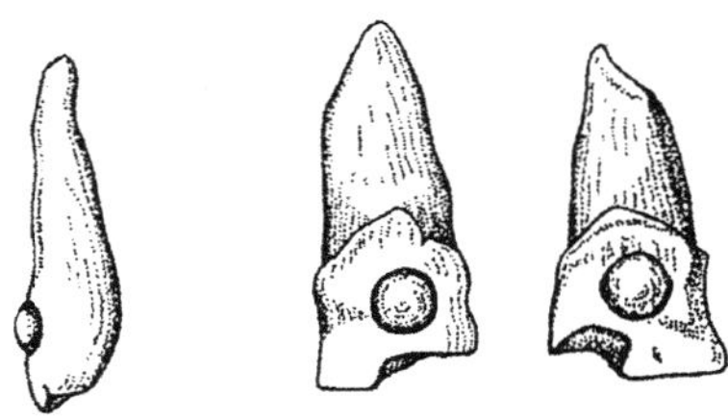

**Fig. 6.** Filed teeth with inlays. Ancient Maya. Illustration by Matilda Essig. Adapted from J. Romero Molina, "Nuevos datos sobre la mutilación dentaria en Mesoamérica," *Anales de Anthropología* 23 (1986): 239.

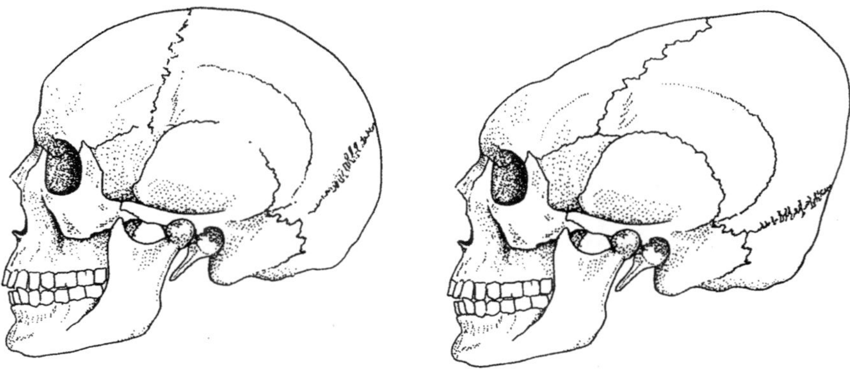

**Fig. 7.** Normal skull and intentionally deformed skull. Illustration by Matilda Essig.

orbital ridge (Dingwell 1931), deform the cranial base, and result in headaches (McNeill and Newton 1965). Finally, and of more significance to Miller's sexual selection hypothesis, is the fact that traditional tribal and clan decoration, one's ancestral visual art, rather than increasing the number of females that a decorated male could attract, as one would predict from sexual selection theory, would limit the number. Given trial endogamy and clan exogamy, a male's decoration would deter women of another tribe or of prohibited clans.

To describe some costs associated with traditional visual art, I turn to the Pascua rituals of the Yaqui of the Sonoran Desert. Those who participate in the rituals are Yaqui, codescendants of Yaqui ancestors. Getting ready to perform the rituals involves making visual art objects that resemble those produced by their ancestors. The ritual is performed more or less as the ancestors performed it. The Yaqui apparently were dancers long before the Jesuit missionaries influenced them around 1617 to blend their ancient traditions with Catholic rituals. These blended rituals continue, and, at the ancestors' request, parents teach their offspring how to perform these rituals and remind them that ancestors want them performed. Masks and outfits resembling those used in the past are made; musical instruments must be tuned; participants must learn to synchronize their movements and the sounds produced by musical instruments. Then, when the rituals begin, men play music and dance for hours on end in the hot Sonoran sun while weighted down with heavy costumes and masks. At ritual's end, costumes and masks are piled in the center of the plaza and burned.

Performing the ritual involves cooperation, synchronization, hours of hard work, and physical discomfort, and it requires the use of a significant amount of resources. Further, males who are not respectful of the ancestors and who have obviously attempted to use the ritual and its art for personal

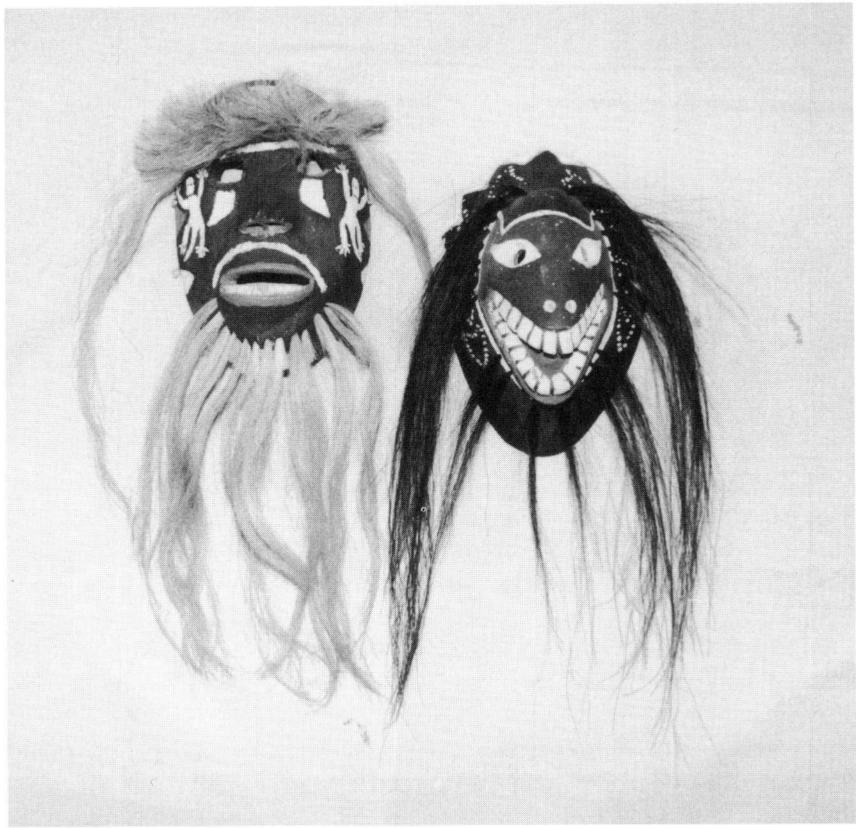

**Fig. 8.** Mayo Pascola *(left)* and Yaqui *(right)* ceremonial clown masks carved in wood with the face of a man and a goat. Photograph by Douglas Mehaffey.

gain or to attract undue attention to themselves as individuals and not as representatives of ancestral figures (that is, through exaggerated movements, sounds, or costume) are criticized. Their participation in future rituals is jeopardized.

Despite these costs, Yaqui rituals are not new; they have been transmitted from one generation to the next for centuries. Males do not seem to be using this ritual and its associated decoration for self-aggrandizement or sexual advertising. Clearly more is going on than kinship behavior; as these individuals share common descent, many are only distantly related. We cannot explain this as group behavior. The Yaqui do not live in a group. They come from near and far. One effect of the visual art and its associated ritual is the communication of shared ancestry; another effect is the encouragement of cooperation, generosity, and sacrifice (of comfort and resources) among the codescendants. It is this effect that promotes the protection, feeding, and education of costly human children, who, along with their parents, would be seen as one's kin and thus the dynastic success of the ancestors who practiced the rituals and asked their descendants to copy them.

While we might be tempted to dismiss these traditional or tribal arts, arguing, as have scholars in the past, that these individuals had no choice in the matter, there are reasons for not being so quick to dismiss traditional visual art. Most important, if we looked over the whole of human cultural behavior, we would undoubtedly find more behaviors have been traditional than have been idiosyncratic. Further, not only do individuals have choices, but also leadership among many of these people was informal. Elders, while enjoying respect, had little power. Finally, close kin, who should be the ones most likely to promote the well-being of their children and kinsmen and kinswomen, are the ones encouraging the sacrifices. Rather than dismissing traditions, we should appreciate them as being of considerable theoretical interest.

## Underpinnings of the Ancestress Hypothesis

The maternal care implied by the ancestress hypothesis is hardly surprising. Like other mammals, we nurse our young, often transmit knowledge to them, and tend to be highly social. We identify kin and preferentially cooperate with them and, like some other primates, we often organize ourselves socially with others in our lineage who share descent from a common ancestor. To some degree, however, modern humans are unique not only when compared with other mammalian and primate species but also when compared with early hominids. Sexual dimorphism has decreased since early hominids, while brain size has increased significantly. Females conceal ovulation and when compared with other primates delay initiating reproduction and terminate it early, long before their death. Females, especially during these extended infertile periods, often serve as allomothers, helping kinswomen rear offspring. Human infants are altricial and remain dependent upon their parents for many years. Males and females form enduring social relationships that involve a sexual relationship and the protection and provisioning of children. Further, humans live longer and during their lives invest a significant amount of resources in complex cultural behaviors.

While hominid females undoubtedly had a variety of mothering strategies, we apparently are the descendants of mothers who selected an extreme K-strategy over an r-strategy. Our species is the most K-selected of all animals as measured by a number of morphological, physiological, and behavioral categories (Bereczkel 1993). As Hrdy explains, "A critical distinction between humans and other animals . . . is the sheer duration and extent of parental investment" (1999, 177). The ancestress hypothesis proposes that the initial development of culture, and a driving force behind the evolution of much of our biology, including our big brain, was this increasingly large investment.

As the human female's investment in offspring increased, the total number of offspring she could produce decreased. As the survival and future reproductive success of each surviving offspring became increasingly more important, our ancestress apparently faced a situation in which loss of an offspring from predators, conspecific males and females, and accidents posed regular threats. Our ancestress protected her costly and vulnerable offspring and prepared them for life, including situations they would face after her death, by building on the maternal kinship and descent (or lineage) strategies seen in other primates and by using the influence that mammalian mothers have over their children. This influence made it possible to develop and maintain the social relationships and division of labor that would be necessary for the honing of cultural strategies, traditions, to solve practical and social problems that regularly presented themselves and to transmit these strategies to succeeding generations.

As primates seem to be able to identify kinship and descent, I assume that early humans could have done so also. By creating mechanisms for identifying individuals who shared common descent, mothers could push the common ancestor back beyond the mother and grandmother identified by some other primates. As the common ancestor became more distant, more and more codescendants could be identified. This is perhaps what E. B. Tylor meant when he argued that the "natural way in which a tribe is formed is from a family which in time increases" (1960, 249). Anthropologists refer to individuals who claim that they share common descent using such terms as tribes, ayllu, sibs, phratries subsections, moieties, and clans. Kinship-like generosity in these categories is so common in the ethnographic record that Fortes (1969) referred to it as the axiom of kinship amity.

Male involvement in kinship and descent relationships, as occurs in some primates, would have been a consequence initially of the influence that mothers had over their sons. The sons of Flo, the dynast chimpanzee from Gombe (see Hrdy 1999), helped protect their siblings. If a male cares for his siblings (since they probably have different fathers, they would be half, not full, siblings), investing his time and energy in their survival and potential future reproduction, he helps fewer copies of his genes move forward than if he had invested those resources in his own immediate reproduction. Each one of a male's own offspring would receive 50 percent of his genes. If he helped his half siblings, or even the offspring of his half siblings, far fewer of his genes (25 percent and 12.5 percent) would be transmitted into the future than if he had reproduced.

Once male primates began investing in their costly younger siblings, as happened with Flo, and helping care for their sisters' offspring, an enduring union between a male and female would eventually come to make biological sense. Although males may produce more offspring by adopting an r-strategy, once our ancestress adopted an extreme K-strategy, a male's dynastic (not reproductive) success would become dependent upon his sexual access to a female with K-strategy skills. The price of access, as offspring represent such a large percentage of a female's total possible reproduction, involved using aggression in appropriate ways—that is, in defense of kin including offspring. At that point, a compromise might become less costly for a male; concealed ovulation would have made this compromise even more intriguing, just as it would have made an enduring relationship necessary.

With the advent of marriage, or an enduring relationship between a male and female, several things could begin to occur. First, marriage could not only significantly increase the number of offspring that a female could produce but also promote their health and well-being (Blake 1955; Wright et al. 1997). Competition would occur between mothers with male kin helpers and social fathers as mates and mothers without such helpers. Competition between males would involve competition not only for K-strategy females but also paternal strategies with a differential effect on the survival and future reproductive success of offspring and, apparently, descendants.

Second, marriage would allow for the identification of paternity, which would have significantly increased the number of kin one could identify. Without marriage, a female could identify only her own offspring or the offspring (female or male) of her female kin (that is, male siblings, mother's brothers, or the male offspring of a sister or other close female relative). With marriage, a child would acquire not only a father but also all of the father's kin (male and female) and all the offspring of all of those kin and the offspring of the males who were their mother's kin. Without the identification of paternity, none of these kin would be identified.

Third, as marriage was endogamous (within the tribe or category of individuals sharing a common, distant ancestor), it basically involved a relationship between codescendants and their kin. A female, thus, would always live among individuals who were her codescendants, shared her traditions, and saw themselves as her kin regardless of whether her residence is patri- or matrilocal.

Finally, another effect of such ties would be that it is possible for a male not only to protect his offspring (which is less likely to occur given polygyny or serial monogamy), but also to address outside threats more rapidly and effectively, as some level of cooperation would have occurred between male kin before any particular threat arose. Loyalty to one's kin would mean that the fighting force would be formidable; the same reactions that prepare mammals to fight or flee are aroused when higher primates perceive a threat to important social relationships (Hamburg 1952, 1963). The nature of the transmission of traditions, from one generation of kin to the next, means that stories of slight, insults, and enemies, just as stories of heroic ancestors, can long endure.

Although a subsistence activity (for example, digging for tubers) may have been the driving force behind a greater maternal and grandmaternal involvement, the strategies that I am describing are more complex and would endure despite environmental change and related changes in subsistence strategies and economic systems. In Australia, traditions persisted for thousands of years, yet it has not been possible to establish a close relation between ecological zone and distinctive cultural elements (Tonkinson 1978, 3). Once complex social strategies, acquired through trial-and-error learning, were in place and were crucial factors in promoting the survival and well-being and future reproductive success of offspring, the failure to replicate these strategies could have long-term consequences.

## The Ancestress Hypothesis and the Visual Arts

If a mother modified the appearance of all of her children, perhaps through a particular hair arrangement or head ornament (such as is seen in the Venus figurines, dating back at least to 28,000–24,000 B.C.E.), or the use of pigments on the body, or even a permanent form of decoration (such as the dental ablation and cranial deformation that were becoming widespread during the Upper Paleolithic), then all her children would be identified as her descendants and as kin to one another. If all (or even most) of her children not only copied her behavior but also, using the influence implicit in the mother-child relationship, encouraged their own children to copy the behavior, decoration in the second generation, instead of just identifying a mother and her children, would identify the grandmother; her daughters, who are sisters to one another and now mothers themselves; and their children, who are nieces, nephews, and cousins.

In the next generation, the grandmother, now probably deceased, would be a great-grandmother and her descendants would include individuals who are second cousins, great-nieces, etc. In each generation, if this practice continued, more individuals and more distantly related individuals would be identified as codescendants of the first decorated mother. Mothers, by changing the appearance of their sons so that they resemble their fathers, could use that decoration to promote a male's confidence in paternity and promote paternal care. Decorating offspring to resemble their fathers could also help promote a social relationship between children and their father as well as their father's kin. As elaborate and colorful decoration attracts the attention of everyone, including enemies and predators, I would not expect mothers to be among the more decorated individuals. If modifying the appearance of children so that they resemble their father does indeed promote paternal behavior, it may be important for mothers to forgo elaborate self-decoration in order to encourage patrilineal-associated decoration.

While the identification of kinship and descent is necessary for kinship and descent cooperation to occur, it is not sufficient. In humans, cooperation between close kin or codescendants must be encouraged; cooperation is enculturated. We are taught, often by our mothers, who our kin are and how we must treat them. Our preferential treatment of codescendants, just as our preferential treatment of close kin, is largely a maternal and traditional strategy. The transmission of traditions from one generation to the next requires kinship cooperation. Further, traditional visual art promoted cooperation, as it was used to decorate or attract attention to messages of appropriate behavior.

Although the ancestress hypothesis proposes that visual art had a maternal origin, it did not, needless to say, remain strictly the province of mothers. Allomothers, male kin, and mates also have played important roles in its development. Males not only make visual art and are often the more decorated; they are, at least in Western societies, art's more rewarded practitioners. The issue, however, is not who is making the art or using it to adorn him- or herself, but what the characteristics are of the art that is being made. I refer to visual art as maternal not only because it is conservative and females tend to be the more conservative sex, but also because so much of visual art centers on kinship and descent. Mothers are centers of families; kinship is identified by birth to a particular mother. Further, the production of visual art requires cooperation, and the themes encourage generosity and emphasize the obligations and duties one has to the elders, to the vulnerable, and to one's kin. The maternal-child hierarchy is characterized by the obligations that the one at the top—the mother—has for those beneath her and for those who preceded her. Further, social mechanisms for promoting the persistence of traditions included the universal rule of "be a good mother" (Edel and Edel 1959), as well as rules encouraging good kinship behavior. Finally, the rules governing the arts are characterized by restraints placed on aggressive and competitive behaviors, which, unless contained, are threats to fragile offspring. In addition, skills necessary to produce art often are acquired during childhood, a period during which mothers have a strong influence on children.

Traditions can be modified and lost. What would make traditions vulnerable to loss or extinction would be the failure of one generation of descendants to replicate the strategies of their parents and ancestors. If one generation failed to do so, those traditions, alone and in their holistic link to other cooperative traditions, would be lost. Selfishness and competition between codescendants, particularly males, would be dangerous for other codescendants and a threat to traditions. For these reasons, traditions, including those of visual art, regularly encourage kinship amity and restraint of competition among codescendants.

Visual art, like other communicative behaviors, is used to influence behavior. While it can be used to promote cooperation, visual art also can be used to promote enmity and conflict. Sometimes, perhaps often, visual art does both at the same time; one example would be an ethnic art that identifies insiders and outsiders. Strabo, in the first century, described a Scythian initiation ritual in which young males clothed themselves in wolf skins and danced in a forest clearing. After these rituals, McEvilley (1992) writes, the young males regarded noninitiates as wolves' prey. How many people have been rescued or killed, protected or raped, given succor or starved based on their art, particularly their manner of dress, a style they inherited from their ancestors? Insiders may use certain adjectives, including the word *noble,* to describe their visual art; outsiders often apply quite different adjectives.

I predict that we will be most likely to follow Jared Diamond in referring to visual art as noble when it and activities often associated with it (music, storytelling, moralizing, etc.) is about kinship behavior involving generosity and sacrifice. Examples of ancestral art would include virtually all traditional or ethnographic art, as well as funeral monuments for ancestors, paintings (and now photographs) of family members and ancestors (perhaps found on a family altar), or, in the case of an ancestral religion, the use of objects said to be identical to those used by a distant ancestor. Other examples would include religious art depicting metaphorical ancestors (for example, paintings of the Rainbow Serpent done by the Australian Aborigines) and, among Christians, Mary the Mother, God the Father, Christ the Son. It would also involve stories and depictions of the Corn Mother, Clay Lady, and Pacha Mama, as found in the Americas. It would include patriotic art (George Washington, the father of his country, risking his life to cross the Delaware River). One common denominator linking all of these ancestors is the lesson they offer, namely the sacrifices that they made for their descendants and, by example and implication, the self-sacrificing behavior they expect of their descendants.

When traditional constraints on social behavior disappear, visual art's characteristics will change dramatically, becoming more like the visual arts often appreciated today: creative, expensive, competitive, and highly individualistic. Innovation is rewarded, and works of art deemed worthy by a cadre of critics often end up with high prices, are often placed in galleries or museums in New York or Paris, or are found in prestigious art journals.

Visual art that is anti-ancestors, antitradition, creative, and used competitively to promote the artist's self-interest (at the expense of others) is not an ancestress strategy. It probably is not coincidental that even today politicians and conservative groups (including groups of mothers) regularly lambaste visual art that contradicts traditional values, particularly when it is publicly funded.

## Reconciling the Ancestress Hypothesis with Modern Darwinian Thinking

Before I can conclude this outline of my hypothesis, a final question is how the ancestress hypothesis fits with modern Darwinian thinking. To begin, while the idea will be controversial, as we seem to be advancing in our understanding of human behavior by using quite different methods, I propose a return to the early ethologists' practice of studying behavior. In fact, one of my aims in writing this book was to show what we could accomplish by resurrecting these methods. The term *mate choice,* for example, was defined as patterns of behavior shown by members of one sex that cause them to mate with certain members of the opposite sex but not others (Halliday 1983). Once a trait was defined, hypotheses were formulated to identify its function, as, for example, that females respond selectively to males that court them most vigorously (Halliday 1983). If females rejected such males, we are probably safe in assuming that the trait would be unlikely to persist, transmitted from ancestor to descendant.

In defining terms, ethologists largely ignored subjective experiences, not because thoughts and feelings did not exist but because they were not accessible to scientific analysis (see Griffen 1984). As Halliday recognized in regard to mate choice: "What we may observe at the behavioral level is selective responsiveness of animals to particular stimuli. The mechanism by which the nervous system brings about such selectivity may involve sensory processing at the sense organ level, matching against a centrally located template, or a preference developed through learning. However, for the understanding of the dynamics of mating systems, the precise mechanism involved is irrelevant" (1983, 4). The same could be said for visual art. Visual art involves the behavior of making and of responding to an art object. While the brain, nervous system, and even conscious desires are undoubtedly involved, we can learn something about visual art by focusing, as have ethologists, solely on behavior. I assume that visual art, like behaviors such as mate choice, is observable, can be defined, and may have an observable social effect that influences its own future replication. Although visual art may be much more than what can be identified with the senses, this possibility does not threaten the assumption that we can increase our knowledge by taking such a focus.

The central problem I faced in writing this book, however, is how we might explain the cross-cultural record of individuals who accepted rather than avoided the high costs of art. Why do we so often find this cross-cultural record of individuals who willingly decorated themselves so that they resembled others with whom they shared descent? Why do we see similarities in body decoration rather than a runaway type selection that produces, as in the case of the peacock and Irish elk, ever more elaborate and unwieldy ornamentation? Why do we find that individuals who share common ancestry but may not be related at all (by our current measures of relatedness) are treated preferentially even when those individuals rarely meet? Why don't we find, in sum, individuals who clearly behaved selfishly?

As we, unlike some early anthropologists, cannot dismiss these people as being prelogical (the modern human brain has been around a long time) or explain these behaviors by reference to dominance (there have not always been individuals who had the power to enforce rules), then how do we explain this axiom of kinship amity? Modern Darwinian theory offers us three possible explanations for this behavior: kin selection, reciprocal altruism, and group selection.

Group selection cannot explain this cooperation. Not only is group selection unlikely to be favored by natural selection, but clans, tribes, and moieties are not groups. Nor can we explain this cooperation by means of kin selection; to give two examples, 7,000 Chachi (Coe 1995) and 200,000 Nuer cooperate as kinsmen and women (Evans-Pritchard 1940). Clearly, not all the Chachi are close kin, nor are all of the Nuer. While we might explain this cooperation as an example of reciprocal altruism, it is reciprocal altruism that has certain unique characteristics. First, individuals involved in this cooperative system all claim descent from a common ancestor and, indeed, as far as we know, share actual ancestry. Second, despite the fact that there are no regular interactions between clan members or tribal members (in fact, individuals may rarely meet), the rule of cooperation is asymmetrical: Cooperate with and render service to those who are identified, through such things as tribal outfits, as your kin and who require your assistance. A return on any investment of altruism may never be paid to the altruist or his or her close kin, but to other codescendants of a distant ancestor they all share. Kin selection theory cannot explain this cooperation; individuals are likely to be more distantly related than the .125 coefficient of relatedness that this theory predicts.

One reason we have been unable to account for these behaviors is that we have defined kinship too narrowly and we seem to assume that the identification of kinship (although the mechanism for identification is not always specified) is sufficient to account for kinship cooperation. Kinship in humans, and perhaps other species, seems to be identified in two ways. First, kinship is identified by birth to a particular mother and the male with whom the mother had a prior relationship. This method leads to the identification of close kin. A second way is by identifying individuals who share descent from a common ancestor. In both cases, we learn to identify kin and code-scendants, and cooperation with them must be taught and encouraged.

A second reason we cannot account for these behaviors is that our focus on the behavior of individuals living today seems to have blinded us to the fact that Darwin's theory largely is about persistence, about inheritable traits that work through time transmitted from one generation of kin to the next (Williams 1966). Darwin referred to natural selection as the "preservation of favorable individual differences and variations and the destruction of those which are injurious" (Darwin 1859, 91). Adaptive traits *are* ancestral strategies. K-strategies do not immediately out-reproduce r-strategies, but they must eventually do so or they would not exist.

Another reason we have been unable to account for these behaviors is our focus on both genes and behavior as selfish. Genes are not selfish; as Symons (1979) noted, genes merely influence cells. Further, selfish gene theory does not imply selfish behavior (otherwise it would be the selfish phenotype theory). The aim of behavior is not to behave selfishly but to transmit our genes into future generations. Our ancestors reared costly and vulnerable offspring in a high-risk environment. Social relationships not only promote successful birth outcomes, as I will attempt to demon-strate, but also the reproduction and safety of one's offspring's offspring, on through generations.

Our focus on selfish genes (and assumption that behavior is selfish) has led us to fail to appreciate the competing maternal strategies that drove up the investment made in offspring, and to focus on strategies that are suc-cessful in the short term, in that they lead to conception or even the birth of a child, even though they may not necessarily be successful in the long term; they may not be successful dynastic strategies. I explore the possibility that self-sacrificing behavior, over many generations, may have dynastic consequences.

In sum, although I am not as daring as Miller (1999), who claimed that "sexual selection through mate choice can potentially explain anything you can ever notice about evolved human behaviour" (1999, 80), the ancestress hypothesis not only offers us an explanation for visual art but also helps us understand why kinship and descent have played a central role in human societies. It can also help us understand decreased sexual dimorphism (and the persistence of tool techniques) as related to early male, possibly paternal, care. It also can help us see menopause, adolescent subfecundity, concealed ovulation, the large human brain, and the long life span as one coherent evolutionary strategy facilitating the greater investment in offspring, including the transmission between generations of important skills. It also can help us understand cultural traits that were widespread in the past and common in traditional societies today: the importance of ancestors; the sacredness of mothers; the cross-cultural respect shown for elders; kinship generosity in descent groups; the idea of purity of bloodlines and the passions associated with ethnicity; the function of hierarchies as opposed to pecking orders; and taboos and the encouragement of restraint, including restraint of aggression and sexual behavior. In a word, we will begin to understand not only the function of traditions but also our ancestors and ourselves.

# CHAPTER 2

# Visual Art Techniques and Veneration of the Dead

Around the world, all traditional visual art and perhaps all ethnographic and prehistoric art, have strong ties to ancestors. In Australia, the Walbiri claim that their ancestors living in the *djugurba,* or dream or ancestral time, created all their graphic designs and that all the designs "represent totemic ancestors" (Munn 1973, 1). All art, from the perspective of the Yolngu (an Aboriginal people who live in eastern Arnhem Land), "is an extension of the Ancestral Past into the present and one of the main ways in which ideas or information about the Ancestral Past is transmitted from one human generation to the next" (Morphy 1993, 245).

The art of Southeast Asia is characterized by continuity with the ancestral past, solidarity with living kin, an affirmation of the social, and cultural integrity of future generations (Crystal 1994). In the islands of the Pacific, all objects of ritual importance commemorated the ancestors. In Central America, "ancestor imagery" was the force "behind the vast majority of ancient Maya public art" (Stuart 1988, 221). In the northwest coast, every object of ritual importance (for example, canoes, boxes, bowls, houses, poles, chairs, clothing, spoons, bracelets) is "decorated with images of the clan's mythic ancestors, with depictions of incidents in the clan's history . . . In some tribes, even people's bodies were tattooed with images of the clan's ancestral spirit being" (Walens 1993, 89). Visual art, in two general ways, is ancestral art. First, art styles are inherited from one's ancestors. One must earn, however, the right to use those techniques, forms, and motifs. The transmission of techniques provides elders with the opportunity to influence the next generation. Second, there is a strong association between visual art and the dead. Bodies of the living are adorned during mourning, often to resemble the dead or his or her totemic ancestor; the bodies and bones of the dead are decorated; the image of the dead is painted or carved; the activities of the dead are depicted; and the burial place of the dead is identified often by using visual art.

There is a close tie between the sacred, or religion; or, as Stuart (1988, 221) expresses it, the "cosmological framework," is closely tied to production of ancestor images. What this means is that the process of painting ancestor art often was equated with performing sacred or religious ritual. Ancestor worship or veneration is a widespread tradition. It perhaps once was universal (Steadman, Palmer, and Tilly 1996), and it regularly involves the claim that ancestors are interested in and interact with the living. One thing that ancestors frequently are said to expect is that their descendants commemorate, in visual art form, their ancestors.

So important and powerful are ancestors and the ancestral art that represents them, one of the worst things that could possibly occur is the desecration of the bodies, bones, tombs, or art of the ancestors. During Chinese interlineage wars, the "surest way to destroy a rival for good is to tear open his ancestral tomb and pulverize the bones they contain" (Freedman 1966, 139). In China, ancestral bones were bones of contention in cases of conflict; pillaging ancestral shrines of rebel groups was a strategy used to crush uprisings (Feuchtwang 1974). The Inca kept the elegantly clad, decorated, and mummified bodies of the dead ancestors among the living and regularly consulted them when important decisions had to be made. If enemies captured the mummies, the ayllu (made up of living codescendants) was considered to be captive (Moseley 1992).

Ancestral icons were regarded as so important among the highland Maya that Spanish efforts to destroy the icons were resisted even though the Maya were tortured and killed by the Spanish when they tried to root out the practice (McAnany 1995, 27–28). Spencer and Gillen write that among the Australian Aborigines they witnessed an occasion in which the *churinga* (which symbolized the ancestors) were removed by a white man who did not understand their significance. "For two weeks, the natives remained in camp weeping and mourning and plastering themselves over with pipe-clay, the emblem of mourning for the dead" (1912, 208–9).

## What Is Technique?

Although we frequently talk about artistic freedom, the production of a work of art depends upon techniques, which are defined in *Webster's New World Dictionary* as the following: "1. The method of procedure (with reference to practical or formal details) in rendering an artistic work. 2. The degree of expertness in following this." While individuals can and do develop and hone techniques, and techniques can remain idiosyncratic, that is used only by one person, techniques also can be taught to others or copied by one's peers, even without anyone being aware of the copying. In kinship-based societies, such as those in which our distant ancestors lived, the transmission of techniques, however, typically involves vertical transmission from ancestors to descendants, often parent to child. Persistence of a technique in the ethnographic and archaeological records thus implies social relationships. The more complex the technique, the longer a social relationship has to endure or the more intense the teaching methods need to be. As methods of transmission are informal in the hunter-gatherer societies that closely approximate those of our ancestors, the transmission of techniques among our ancestors more than likely involved time, not intensity.

Anthropologists assume that the refinement of technique requires a particular kind of social relationship, namely one involving a division of labor, which, as Franz Boas argued, was a defining feature of visual art. In chimpanzees we see a division of labor of sorts. When a mother transmits techniques to her offspring, the mother leads; the child follows. Teaching, in sum, is a division of labor. The offspring watches, copying the mother until some level of skill is acquired. Further, in the primate mother-child relationship we also may find the rudiments of a sexual division of labor. In Gombe, Flo had her sons protect their siblings while she performed other activities.

While it is not clear when a division of labor began to play a role in the human production of art objects, by the Aurignacian (about 35,000–30,000 B.C.E.) and the Gravettian (28,000–22,000 B.C.E.), techniques had become quite complex. Leroi-Gourhan, who coined the words *chaînes opératoire*, used them to describe the complicated techniques that involved habitual, learned sequences of technical operations, or standardized movements, and the production and use of standardized tools (for example, adzes, scrapers, and drills) to produce objects (1943). Teaching is a division of labor that can lead to standardization. As these techniques are complex and learned, they imply social relationships that endured for some time. Anderson argues that a sexual division of labor is crucial to the development of technical competence: "the technical excellence that is apparent in Northwest Coast carving and weaving is contingent upon the sexual division of labor whereby a woman can devote all her time and energy to mastering weaving, and a man to improving his skills of carving" (1979, 87). The fact that human sexual division of labor is now found, as Anderson points out, "in virtually every society's art" (85) suggests that it is ancient and important. The way the sexual division of labor is expressed suggests it is influenced significantly by our genetic inheritance. Across culture, men tend to work with hard materials (for example, stone, horn, shell), while women tend to work with softer materials such as clay, fiber, and leather. There are, of course, exceptions. Among most North American tribes, women did the tanning, while in South America (except for Ona women), it was a man's job. Hopi men spin and weave, jobs that the Navajo allot to women (Lowie 1940, 108).

Although a division of labor can facilitate the development of virtuosity, a division of labor, however, also has costs. The players become vulnerable; an offspring's long-term survival, for example, depends upon mothers who provide them the opportunity to copy. Females must depend on males to produce what they do not, just as males must depend upon females. If one person fails to do his or her job, all will pay the consequences. A division of labor also limits one's ability to do what one wants to do or even what one might do best. What a division of labor may do is promote trusting, enduring social relationships among the generations and sexes.

## Technique: How Is It Acquired?

While we talk about an association between technical competence and teaching, the teaching need not be formal. Parezo describes how the Navajo learn to do sandpaintings: "Teaching was by showing, without explanation or generalization. Often no instruction was given, nor, as one sandpainter said, should it be, for learning is in the environment; it is never conscious. Most sandpainters do not think they have been taught, because they associate the word 'teach' with a classroom situation" (1983, 122).

In Ecuador's coastal rainforest, Chachi girls have no formal training or lessons. Technical skills were acquired gradually, through constant observation of their mothers at work. It took a girl approximately twelve years to learn to weave a traditional basket, as she had to learn where to find the raw materials and how to process and prepare them for weaving. She learned to weave by observation, beginning when she was an infant sitting on her mother's lap watching while her mother made a basket. During these years, she also learned how a Chachi woman was expected to behave (Coe 1995).

Informal transmission of behavior, as these examples imply, can be time consuming. Among Australian Aborigines, men learn painting techniques from older male kin during an initiation process that may take up to ten years. A benefit of this slow style of learning is that students acquire skills as they are ready, physically, mentally, and emotionally, to acquire them. One is taught a skill when one shows one's readiness. The transmission of the most restricted and sacred clan paintings occurs after a male has mastered all other techniques and shows he is ready to be trusted with the sacred. This generally is "shortly before or soon after their father's death. If a man dies when all his children are too young to be taught the paintings, the responsibility is passed on to someone else [in the clan]. In explaining why they taught the children of a dead man the paintings belonging to his clan, people often alluded to deathbed conversations in which the dying man entrusted them with the responsibility of teaching his children the clan's *mardayin* at the appropriate time" (Morphy 1991, 63).

While parents and grandparents have the obligation to transmit art behavior to the next generation, children have an obligation to acquire ancestral ways and to behave respectfully around the elders, who are the holders of traditional knowledge. Spencer and Gillen (1938) write that among the tribes of central Australia, a male is said to have completed his initiation

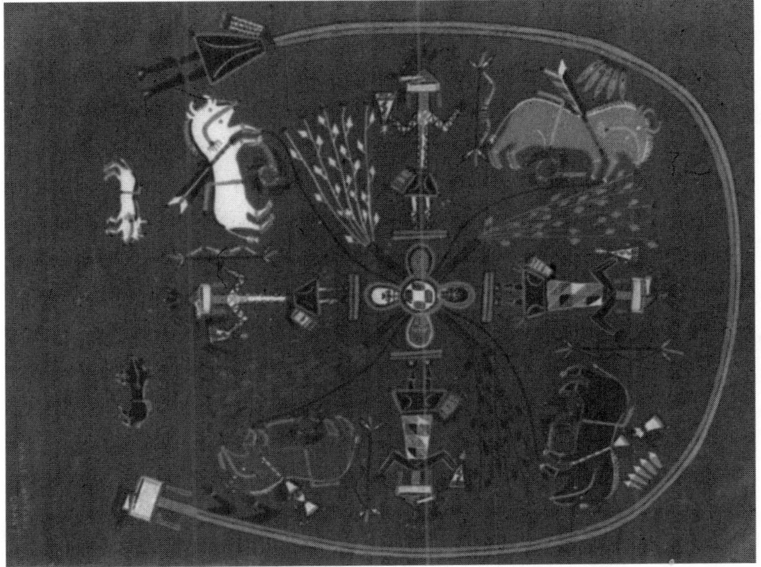

**Fig. 9.** Navajo sandpainting. Courtesy of the Arizona State Museum, University of Arizona.

only when he has learned not only his ancestor's traditions but also respect for the elders who are the living ancestors.

Anna Shaw, a Pima woman of Arizona, described how her grandmother taught her to weave. Anna spent hours gathering black devil's claw, willow twigs, and cattail reeds with her grandmother as they walked together across the Sonoran Desert. By watching and helping, Anna learned when and how to gather the materials she would need. By helping her grandmother process these materials to get them ready for weaving, Anna developed her own skills. Watching her grandmother weave, Anna saw the basket forms grow and the design motifs emerge. Her young fingers found weaving difficult at first, and baskets had to be woven, then unraveled and rewoven, until the weaving was properly done. She tells us that her grandmother told her "basket-weaving is to teach you patience, my granddaughter" (1974, 50). When her first well-made basket was completed, she was considered a woman; her adult status began when she gave this basket to her grandmother.

What Anna Shaw's story suggests is that the process of transmitting the skills necessary to make a basket is important, just as are the products of that process, and that acquiring these skills, largely through observational learning and imitation, is a lengthy process. Those years, however, also provided an opportunity for a girl to develop a relationship with her grandmother

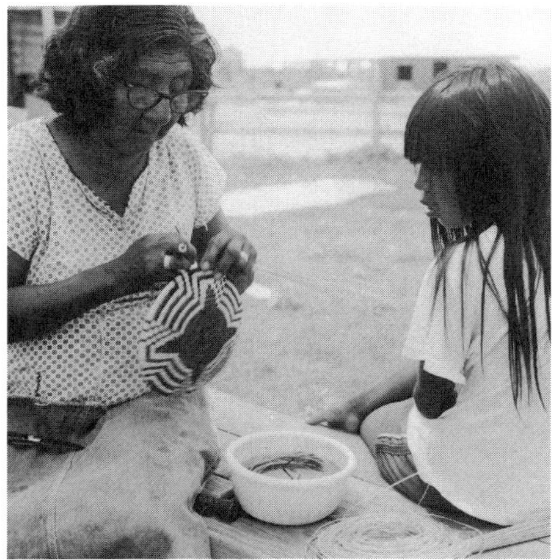

**Fig. 10.** Pima grandmother showing how to weave a Pima basket. Courtesy of the Arizona State Museum, University of Arizona. Photograph by Helga Heiwes.

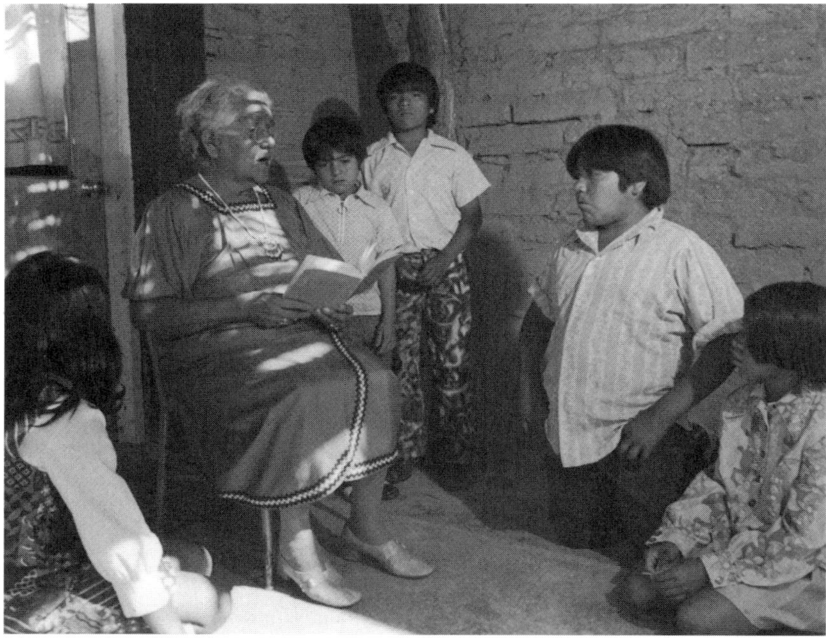

**Fig. 11.** Anna Shaw reading her book, *A Pima Past,* to children and grandchildren. Courtesy of the Arizona State Museum, University of Arizona.

and hear stories about the ancestors they shared, including the meaning of the design motifs they both had inherited from those ancestors, back to the first Pima women who had woven the first baskets. The ancestors expected the grandmother to teach her granddaughter to weave. They also expected, Anna was told, the granddaughter to learn to weave, to show her generosity by giving her first well-woven basket to her grandmother from whom her weaving skills came, and, down the line, to teach her own daughters to weave. They also expected her to behave respectfully around the elders; to be loyal and generous; to be a good wife and a good mother; and to cooperate with and be generous to her kin, first her family and then the other Pima. Other Pima were those with whom Anna shared a common ancestor and who could be so identified not only by their language and manner of dress but also by the distinctive baskets woven by the women.

Traditional visual art links parent and child, ancestors and descendants, kinsmen and women. When kinship and ancestors are strong, tribal arts will be conservative, in the sense of appealing to the traditions of the ancestors. If no other explanation of a practice is available, one can always say "the ancestors did it this way" (Moore 1995, 17).

When I conducted my own fieldwork among the Chachi of Ecuador's lowland rainforest, I was regularly told when I asked someone why they did something in a particular, formulaic way that that was the way the ancestors did it and the way the ancestors wanted it done. Ancestors are not only the ones who gave life but also the ones who gave the descendants their visual art and, along with that art, a blueprint for how to live life successfully.

## Ancestors and Traditional Visual Art

### Mortuary Practices

"Most, if not all, contemporary or historic societies have invested considerable artistic materials and efforts in the disposal of their dead" (Biggs 1989, 1). Not only do humans bury their dead; they have been doing so for thousands of years. One of the oldest burials yet discovered was of an infant buried between 70,000 and 80,000 years ago at Border Cave in southern Africa. The burial was adorned with a perforated *Conus* shell brought from some 80 kilometers away (Grün and Stringer 1991).

Not only is the practice of burying the dead ancient, but specific traditions lasted long periods of time. As Binford pointed out, burials are one of the more conservative traditions (1971). Burial pyramids in Egypt were used for almost five thousand years. Burial practices at Paloma, in the Andes, did not change significantly during a three-thousand-year period (Quilter 1989). While persistence in burial style would seem to imply kinship and descent, the relationship is not always an easy one to prove. One clear exception would be the burials at Broadbeach Cemetery. In the seventh and sixth millennia B.C.E., before the earliest cemeteries, the southeastern European Neolithic cultures of Sesklol, Starcevo, and Karanovo seem to have interred family members, females and children, under the floors of what presumably was their home (Gambutis 1991, 331). The ethnographic record also shows a strong relationship between kinship and burial place and style. The Chimbu of New Guinea had special cemeteries on clan lands for clan and/or subclan members (Brookfield and Brown 1963, 40). The Central Mae Enga buried the dead on clan land, often near a clan forest (Meggitt 1965c). The Kuma had clan burial grounds (Reay 1959), and the Fore buried the dead in the gardens of the lineage (see Binford 1971).

Binford has proposed that burials will be more likely to have a degree of permanence in societies in which ancestry is important (51). In other words, when ancestry is important, the place of burial will be made to endure, unlike the shifting garden burials listed above (which eventually leave no trace). What may be more significant when ancestry is important, however, is not only to acknowledge the dead through art and ritual but also to position the burial not too far from the home, so that burial rituals can continue over the years. The Maya of the Yucatán Peninsula and Guatemalan highlands, for example, "interred their ancestors under the floors of their houses, in residential shrines, and within large funerary pyramids right in the center of their cities and villages." It was important to the Maya to live with the ancestors because rituals for the dead had to continue for years (McAnany 1995, 1).

Among many traditional people, burial rituals are protracted, lasting years, decades, or even centuries. The LaDagaa of Africa had elaborate burial rituals that lasted until the formerly living was transformed, through rituals, into an ancestor (Goody 1962). This process could last some time. Among the Mardujara Aborigines, the widow and other close relatives of the deceased cut their hair and painted themselves with red ochre throughout mourning, a period of one to two years (Tonkinson 1978, 85). The ancient Maya are known for superimposing new pyramids on top of old ones built centuries before. Rather than seeing this as an example of the destruction of the earlier pyramid and its ancestors, it should be seen as remodeling, done in homage to those ancestors (McAnany 1995, 51).

The practice of attending to or feeding the ancestors for years after their death is known around the world. "The attentions of the living to the dead" among the Ma'anyan of Padju Epat, Indonesian Borneo, "do not end with the funeral," as the spirit of the dead has to be guarded and fed periodically until the body is finally cremated (Hudson 1972, 125). In many African kingdoms, females were tomb guardians (Feeley-Harnik 1985). Among the

Maya, feeding the ancestors was an integral part of the domestic ancestral rituals performed at ancestral shrines. Father Diego de Landa (1524–1579) wrote that Maya women "were very devout and pious and also practiced many acts of devotion before their idols [ancestors], burning incense before them and offering them presents of cotton stuffs, of food and drink" (Tozzer 1941, 128). The *huaca* ancestors of the Inca were offered libations of *chicha,* a lightly fermented corn drink.

In China, females were in charge of the domestic commemoration of ancestors (Freedman 1966). In an eighteenth-century Chinese novel, the author described how food offerings were passed from hand to hand by kin until the offerings reached the hands of the oldest family members, who raised the dishes "reverently towards the portraits before laying them down on the altar . . . Meat, vegetables, rice, soup, cakes, wine and tea [were] all transmitted to the alter in this human chain" (Cao Xuegin qtd. in Stuart and Rawski 2001, 47). In China, "it was believed ancestors could bestow upon the living the blessings of longevity, prosperity, and progeny and paying homage to the ancestors by placing food offerings before their portraits was a sacred family duty" (Stuart and Rawski 2001, 6).

In many parts of the world, burials are elaborate. The Yolngu of Australia, as one example, had primary, secondary, and tertiary burial ceremonies, as well as a succession of purification rites that were performed in the months and even years following a death (Morphy 1984, 33). Not only was attending the dying a kinship obligation (most Yolngu by the time they entered their teens had sat for hours at the side of dying relatives, singing through much of the day and night), but all of the deceased's relatives attended some part of the three major funeral ceremonies. Individuals related to the deceased and holding a specific kinship relationship were charged with determining what rituals were to be conducted, as well as when and where they were to be conducted.

Among the Yolngu, the deceased initially was buried on a platform (Morphy 1984). After a period of time, the body was removed from the platform and a secondary burial took place. In this burial, the bones were placed in a painted wooden cylinder, which was sealed and carried around by relatives, frequently for years, until the final burial, which consisted of placing the bones in a hollow log coffin, where they remained. In this final ritual, the deceased was painted with sacred ancestral designs and the same design was painted on the coffin. During the twenty-four-hour period it took to complete the painting, the songs of totemic animals were sung. The Yellow Ochre ceremony, which began soon after the painting was completed, referred to the ancestral beings who had dug up yellow ochre from the Gurunga quarries at Caledon Bay. During this ceremony, the older men painted their bodies with yellow ochre so that they resembled the ancestors.

The younger men, who were not yet worthy of this decoration, were painted in pipe clay.

Egyptian burials often were elaborate. Lucie-Smith described the objects found in the Egyptian burial of the late-eighteenth dynasty pharaoh Tutankhamun (ca. 1361–1352 B.C.E.): "Thrones, beds, stools, chariots, games and game-boxes, caskets for clothes, writing materials, walking sticks, lamps, and musical instruments—and also magical objects and amulets . . . There was an array of beautifully made model boats for his journeys in the after life and a mass of *shabtis* (small figures which would serve as the king's deputies if he were called upon to perform menial tasks in the nether world)" (1992, 39).

Tutankhamun's burial was hidden beneath the larger tomb of Ramses VI (1151–1143 B.C.E.). Tutankhamun's body was well protected: "First there was a series of four gilded shrines in which was a massive yellow quartzite sarcophagus, exquisitely carved with the tutelary goddesses in relief. This sarcophagus contained three coffins, nested one within another. The two outer coffins were of gilded wood; the innermost . . . is made of solid gold. It is richly inlaid with hard stones and colored glass. The king's mummy wore a gold portrait mask covering its whole head and shoulders. This, too, is richly inlaid . . . The effect is startlingly lifelike" (39).

As the investment made in any behavior, including ritual behaviors, is relative to the resources available, both the Yolngu and Egyptian burials should both be considered complex, in the sense that both were time and resource consuming. What is less obvious is the link to kinship and ancestors. In the case of the Yolngu, this tie, however, is obvious. This tie should be equally obvious in the Egyptian burial. Tutankhamun's name means the "living image of the state god Amun." He was a priest king; his authority was said to come from the god whose spirit dwelt within him. The pharaoh, in other words, was the living ancestor.

## *Decoration of Ancestors' Bodies*

Bodies and bones of the dead were decorated in many ways. In the outer islands of Indonesia, the bones of ancestors are curated by family descendants and can command a prominent place within the community (Taylor and Aragon 1991). The Muisca Indians of Colombia were said to have brought their mummified ancestors with them into the battlefields (Cárdenas, in press). The mummified Inca rulers were elaborately dressed and paraded during religious festivities (Moseley 1992).

In some traditional societies, the bones of ancestors were placed in decorated boxes or bundles so that the ancestors could travel with their migrating descendants. The Quiché Maya carried sacred bundles that were

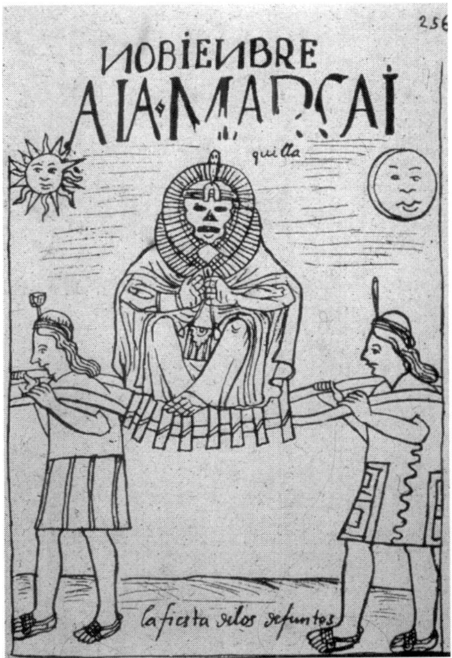


**Fig. 12.** A sacred Inca mummy paraded on a litter in Peru. Facsimile of a drawing by Felipe Guaman Poma from his *El primer nueva cronica y buen gobierno*. Courtesy of Werner Forman Archive/Art Resource, New York.

handed down from the founding lineage ancestors to their successors. These bundles contained relics (crania and long bones, in particular) of the founding ancestors (Carmack 1981, 63). The skull and cross bones, frequently depicted in ancient Mayan art, may date back to the long bones carried in Mayan bundles (see McAnany 1995, 46–47). The practice of carrying ancestral bones may also help explain why bones are regularly missing from Mayan burials.

In New Guinea, skulls were removed from the body after burial. The skull could be dried out on a pole and then placed in the home. Designs could be etched into the bone before it was completely dry or the skull could be covered with clay and decorated to resemble the living person. Kuru, a slow-acting neurological disease, may be a result of the practice of removing and decorating ancestral skulls (Steadman and Merbs 1982).

Long before the Egyptians practiced mummification, the prehistoric Chinchorro people of the coast of Chile developed a complex method of artificially mummifying the dead. While earlier mummies in the area were naturally created by the dry climate, the earliest artificially created mummy

(5050 B.C.E.) was of a child. Mothers may have started the practice of mummification. The earliest mummy is assumed to be a child, because statuette mummies, which are doll-like and contain human fetal bones, are not uncommon in the area (see Arriaza 1995).

The bodies of these children and adults were eviscerated and defleshed, and the skeleton was reassembled, reinforced, and filled with packing materials such as reeds and sea grass. The skin was replaced and covered with an ash paste and then, in the case of the earlier mummies, painted with a coat of shiny black manganese; faces were covered with a clay mask. This process required a great deal of skill and time and increased one's risk for disease, including treponema-like infections. The practice of preparing these black mummies lasted three thousand years and then was replaced by the practice of painting mummies red, which lasted about 500 years (Arriaza 1995, 120).

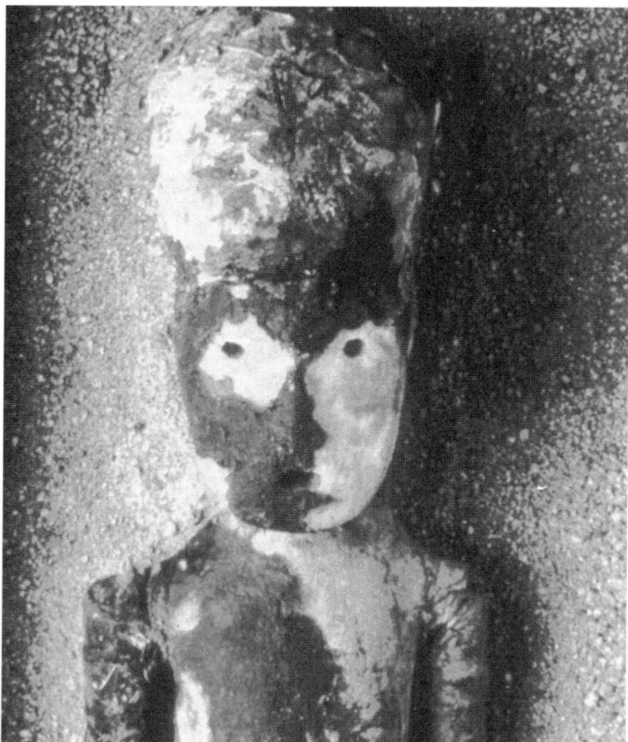

**Fig. 13.** A Chinchorro red-style child mummy from coastal Chile, circa 2000 B.C.E. Photograph by Bernardo Arriaza, Department of Anthropology, University of Nevada, Los Vegas.

Once the mummy was painted, elaborate clay helmets or wigs of human hair were placed upon the head. These mummies were clearly repaired and repainted, probably because they were kept on display and perhaps fed (as were subsequent Inca mummies). Eventually, the black mummies were placed in a mat shroud and buried in the sand in what appears to be family graveyards. All individuals—old and young, men, women, children, and infants—were mummified; there is no evidence of ranking (134). The Chinchorro mummies are so elaborate that scholars have, due to the plasticity of the mummies' shapes, their colors, and the mixed media used in their creation, referred to them as works of art (see Arriaza 1995).

Chinchorro mummification influenced the Inca practice of mummification of ancestors. In Peru, duties associated with ancestor worship included keeping the corpses intact and close at hand. During Inca and pre-Inca times, people consulted and propitiated their progenitors on a regular basis. These forebears defined the lineage, moiety, and ayllu to which an individual belonged. The tradition of mummification ended with the potentates of Chimor and Tahuantinsuyu, whose richly clothed mummies were carefully attended at special shrines. Inca royal mummies, regarded as quasi-alive, were regularly paraded about and formally seated at important council meetings so that they might be consulted and guide the living (Moseley 1992, 53–54).

## Representation of Ancestors and Ancestral Events

The practice of making visual art that depicts the deceased ancestor is widespread. The Toraja of Southeast Asia live in a valley in South Sulawesi, the fourth largest island of the Indonesian archipelago. The funeral rituals conducted by the Toraja are elaborate and complex; buffaloes and pigs are slaughtered and, for at least the past twelve generations, the Toraja have commissioned ancestral statues for the funerals. Not only are statues made for funerals, but they are also regularly refurbished, reclothed, and consecrated. The Toraja have been known to tend to thousands of statues, all of whom are ancestors (Crystal 1994). This tradition of making ancestral statues was built on an even older tradition. In China, we find among the twenty-four paragons of filial piety the account of Ting Lan, who carved an image of his deceased mother and worshiped it (Mayers 1874, 219). Marco Polo reported that in Ceylon, legend had it that the first idol was made after the Buddha's death, when the Buddha's father had a golden image made of his son (Zimmer 1955, 409). Vasari told how Signorelli painted the body of his son "so as, by the agency of his hands, to have as often as he wanted before his eyes one who nature had given and an inimical fate had snatched away" (1986, 691). Among the Kwakiutl clans of the northwest coast of the

United States, ancestors are said to come down to earth in the form of the birds or other animals depicted in the masks and totem poles (White 1993, 288). The totem pole animal images were thus said to be images of the human ancestors.

## CARVINGS OF ANCESTORS

The origin of the practice of carving ancestral figures may be decoration of the actual body or bones of the deceased. Moore (1995) argues that the fact that ancestral skulls are treated with affection and awe in many parts of the world suggests that masks, which often are of ancestral faces, might be a practice that evolved out of the decoration of ancestral skulls. Death masks were made in various places, including Europe, Africa, and South America. These masks were placed on the face of the dead or kept in the home.

In New Guinea, in the areas of the Papuan Gulf, the word *gope* is used to refer to bull-roarers (often referred to as the voice of the ancestor), skull racks (which are made to hold ancestral skulls), and ancestor or spirit boards. The most sacred *gope* boards were clan boards, which were given names. Smaller *gope* boards, which were also named, stood at the entrances to clan areas in the long houses. The small *gope* boards, which were unnamed, were given to boys and young uninitiated men. These boards were hung in the sleeping area, as they were said to help males grow strong.

Korwar figures, which were carved in northwest New Guinea, were sculptured squatting figures. The word *korwar* refers to the spirits of the dead; however, the term also is extended to the skull of the ancestor and to the wooden ancestor korwar figures that often contain an ancestor's skull (Baaren 1968, 85). Figures were carved during a person's lifetime so that at the moment of that person's death, the spirit could enter it (Linton and Wingert 1946, 134). As the carved figure represented an honored ancestor, and often contained the bones of that honored ancestor, the family kept the figures in their house to serve as mediators between the living and the dead.

In the traditional art of the Asmat, who inhabit a vast swamp on the south coast of New Guinea in the Indonesian province of Irian Jaya, objects are named for an ancestor both in dedication to that ancestor and to seek that ancestor's protection and goodwill. Ancestors are represented on or associated with all Asmat objects (bowls, pipes, masks, ancestor poles, spears), including war shields. These war shields were once divided into four categories based on their characteristics. Today, however, twelve distinct cultural categories have been identified based not only on the characteristics of shape and iconography but also shared language, myths, feasts, etc. (Schneebaum 1985; Smidt 1993).

In Vanuatu (New Hebrides), statues that represented the recently deceased ancestor were carved from tree-fern trunks and teak wood and used

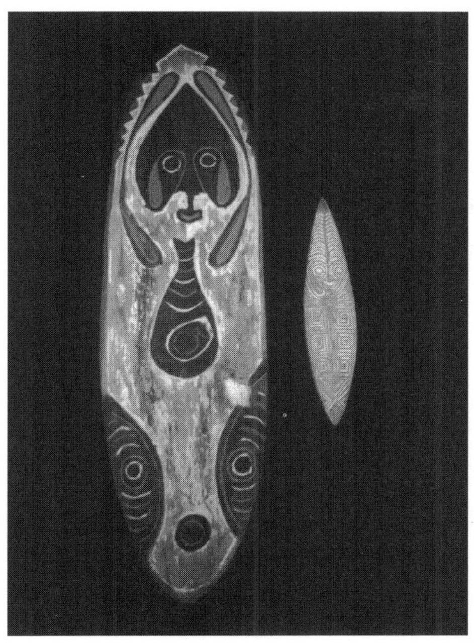

**Fig. 14.** *Gope* ancestor board, Papuan Gulf, Papua New Guinea, Kerewa area. *Left,* initiates board from Goaribari Island (wood, lime, red ocher, and charcoal). *Right,* small, unnamed *gope* (wood and lime). Photograph by Douglas Mehaffey.

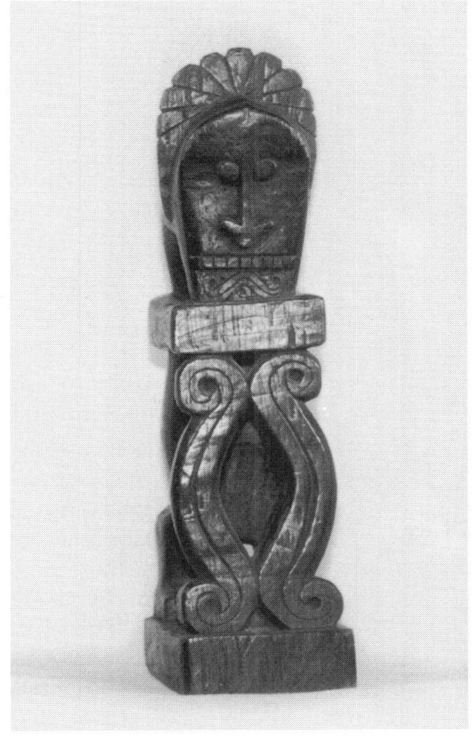

**Fig. 15.** Contemporary Korwar figure. Photograph by Douglas Mehaffey.

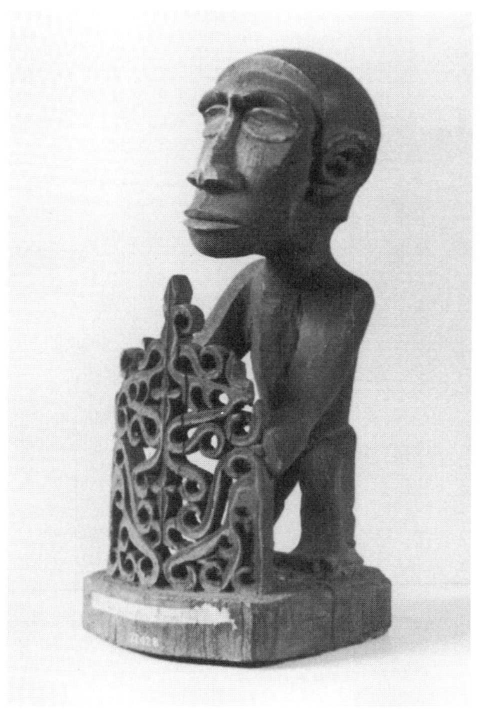

**Fig. 16.** Mortuary figure from Dutch New Guinea. Courtesy of the University of Pennsylvania Museum, Philadelphia.

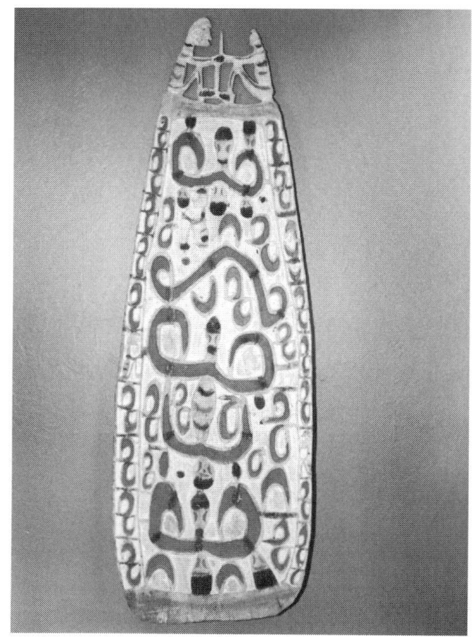

**Fig. 17.** Contemporary Asmat shield from Irian Jaya, Indonesia, Sawa-Erma, Pomatsj River, Northwest Asmat (wood, lime, red ochre, and charcoal). The primary motif is the flying fox figure. Ancestor figures appear at the top of the shield. Photograph by Douglas Mehaffey.

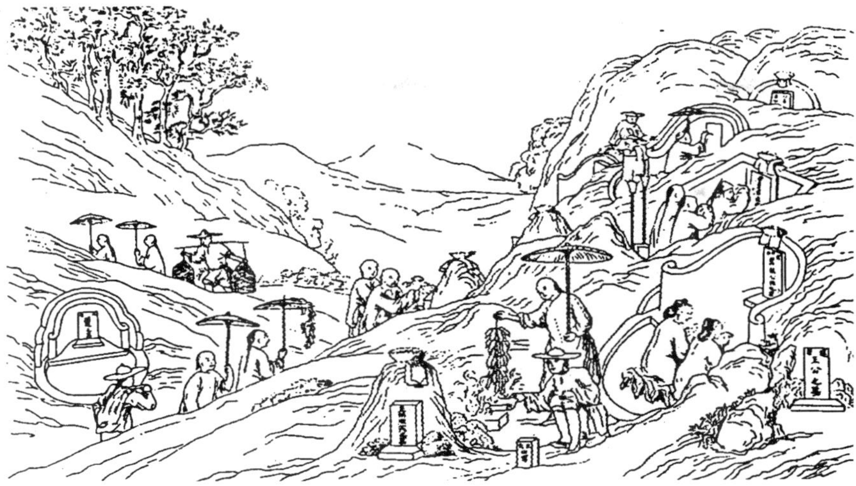

**Fig. 18.** Tombs of the ancestors in the countryside of China. Adapted from M. Freedman, *Chinese Lineage and Society* (New York: Athlone Press, 1966).

in mortuary ceremonies. Once carved, these statues were covered with a clay coating and painted in earth pigments of bright colors (Moore 1989). In the Marquesas Islands, figures called tiki were carved from wood or stone. The sunken eyes and shrunken lips of the tiki figures may be related to the island's cult of the dead, which had at one time involved mummifying one's ancestors (Linton and Wingert 1946).

Tongan ancestral female figures were carved of whale-tooth ivory and worn as pendants. These female ancestors served as models and archetypes for women (Moore 1995). When girls wore these pendants they were reminded of appropriate social behavior.

In the New Territories of China, ancestral paintings depicted the ancestors and certain ancestral events. Tablets, which were commemorative of both male and female ancestors, were a crucial part of domestic shrines (Freeman 1970). The burial site of the ancestors was marked by a tomb, which was visited and carefully maintained.

In Africa, the production of ancestral figures was sometimes affected by inhibitions about the realistic portrayal of the human form. Yet the Ambete, who claim a Kota origin and who live in the part of the Republic of Congo near the northern frontier of Gabon, carved heads, busts, and figures. The Ambete practiced ancestor worship, and their carvings are connected to the ancestor's cult. The full figure carvings were used either as reliquaries or placed alongside ancestor bones in a basket. Sacred relics, such as ancestor

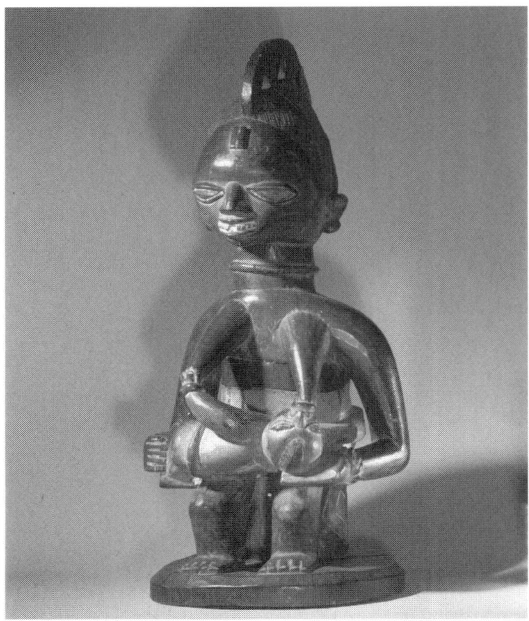

**Fig. 19.** A wooden maternity figure from Yoruba, Nigeria, probably given as a thanksgiving offering to an *orisha* shrine following successful prayers for a child. Courtesy of the P. Goldman Collection, London, and the Werner Forman Archive/Art Resource, New York.

bones, were placed inside the statues, which had hollowed-out backs. The carved heads and busts were placed outside on poles.

In their second burial ceremony, the Nigerian Owo, Benin, and Onitsha people made a figure referred to as *ako,* which represented the dead (Willett 1971, 360). Among the Vatshiokwe in West Africa, masks are named after the dead and resemble them. Baumann reported that "I heard time and again the masks are identical with the deceased ones" (1935, 105). Among the Don of West Africa, making a mask requires "the assistance of the ancestors. The Don sculptors claim that ancestors help only if they abstain from sexual relations before beginning the work" (369). Ancestral masks, the highest or most sacred masks, are used in rituals at which families make sacrifices to their ancestors (Gerbrands 1971).

Funerary stela were made in Angola (Solongo people) to commemorate women (Blier 1998, 220). Wood sculptures of women nursing or holding a child, which were carved by the Kongo people of Africa, represented women as founders of the dynasty. The theme of these wooden sculptures

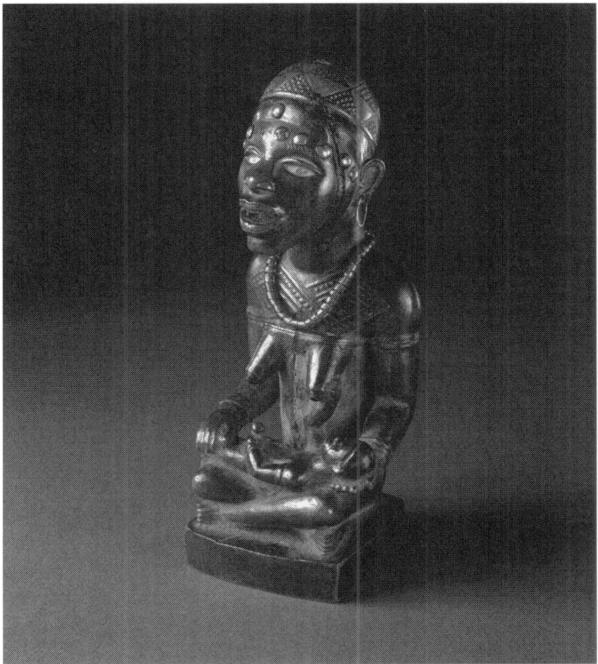

**Fig. 20.** Woman and child figure from Congo and Zaire (wood, mirror, glass beads, metal, and brass tacks). Courtesy of the National Museum of African Art, Smithsonian Institution, Washington D.C. and Aldo Tutino/Art Resource.

was to point out the role of women in "assuring continuity of the lineage" (Blier 1998, 221).

The Poro society of Kufuro villages on the Ivory Coast is "above all an organization designed to maintain the right relationship with the Deity and the Ancestors. The most important ancestor is the woman who was head of the founding matrilineage of each *sinzinga*" (Glaze 1993, 222). Poro rituals are under the authority of Ancient Woman. Her art was used to maintain social order by means of a complex system of education, which was designed to help shape the intellect, the moral character, and the skills of each succeeding generation and age set (224).

On the northwest coast, images carved on totem poles represented the ancestors. Further, the Northern Kwakiutl people carved grave effigies probably intended to be portraits of the deceased. A mask representing the face of an old woman's spirit was found in a shaman's burial house of the Tlingit, perhaps indicating that it was considered to be of powerful ritual use.

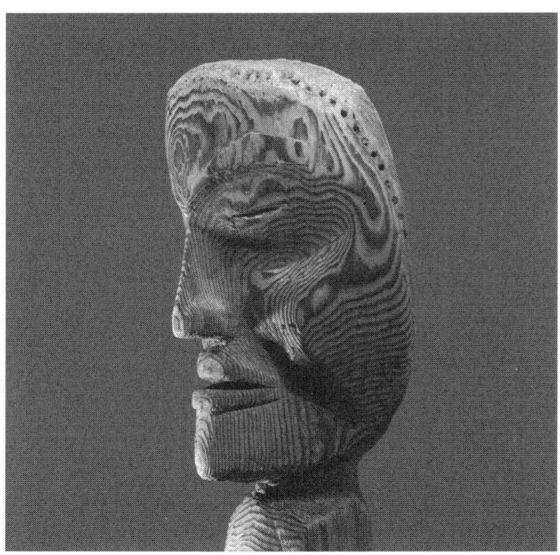

**Fig. 21.** Grave effigy of the northern Kwakuitl people, possibly intended as a portrait of the deceased (red cedar wood). Courtesy of the Museum of Anthropology, University of British Columbia and the Werner Forman Archive/Art Resource, New York.

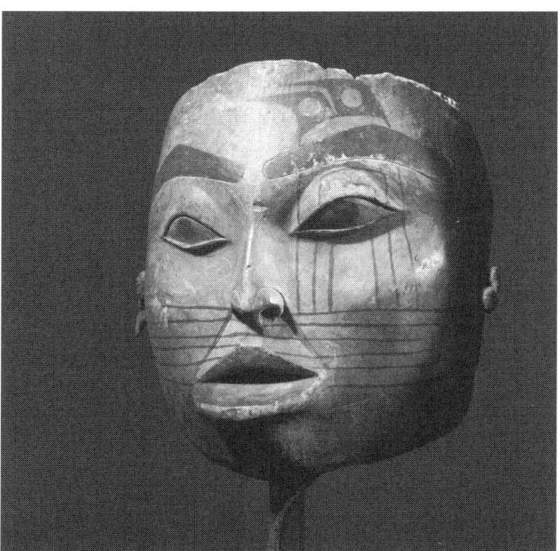

**Fig. 22.** A Tlingit Indian mask representing the face of an old woman's spirit. Found in a shaman's burial house on the northwest coast of North America. Courtesy of the Field Museum of Natural History, Chicago, and the Werner Forman Archive/Art Resource, New York.

## ANCESTOR HOUSE DECORATION

In the Solomon Islands, representatives of the ancestors were carved into posts cut from tree trunks. These posts provided the central support of houses, just as the ancestor provided the central support of the family. In New Caledonia, large figures carved from tree trunks were set up on the dance grounds to honor the ancestors. These male and female figures, which were said to be the guardian ancestor spirits, were almost human in size (Linton and Wingert 1946). In the Sepic, the Arambak people carved figures of their ancestors from wood pieces left over from a drum made from a tree growing at the grave of a female ancestor (Dennett 1975; Forge 1973). Further, painted faces *(ngwalndu)* were displayed in rows on the exterior facades of houses. The big eyes on the faces, Forge explains, are the eyes of the ancestors who have come back to observe their descendants' behavior.

Among the Maori, an ancestor figure was carved after death and placed in the central house of the chief meeting room. This figure was said to bring mana, which is a spiritual power coming through a chain of ancestors and manifested in chiefs, their descendants (Moore 1989). The Pukeroa gateway, which was constructed in 1836 for Pukeroa pa, Ohinemutu, was known as "Tiki." This gateway depicts an ancestor, Tutanekai, who bears the tattoos of his ancestors; in Arawa moko (or tattoo), the father's ancestors are depicted on the right-hand side of the face and the mother's ancestors are on the left.

## PAINTINGS OF ANCESTORS

The origin of paintings that depict ancestors may have been the practice of painting and decorating the body of a living descendant so that that he or she resembles a particular ancestor. When a male decorates his own body or paints or engraves sacred symbols, he is said to become "the hero or the ancestor. By reenacting what the ancestor did, he becomes a life-giver too" (Elkin and Berndt 1950, 2). The "painting of ancestral designs on a person's body is a means of directly associating that person with the ancestral worlds" (Morphy 1991, 131).

In imperial China, all social classes considered it a sacred family duty to honor the spirits of their deceased ancestors in ritual ceremonies (Stuart and Rawski 2001). Portraits of ancestors, including the father and mother, were painted and brought out for rituals. As the ancestors were said to continue to be influential, appeasing them by worshiping their portraits affected their descendants' wealth and the number of progeny. All ancestors were painted with virtually the same somber and detached expression. They were painted to look withdrawn, as though they were ready to face eternity

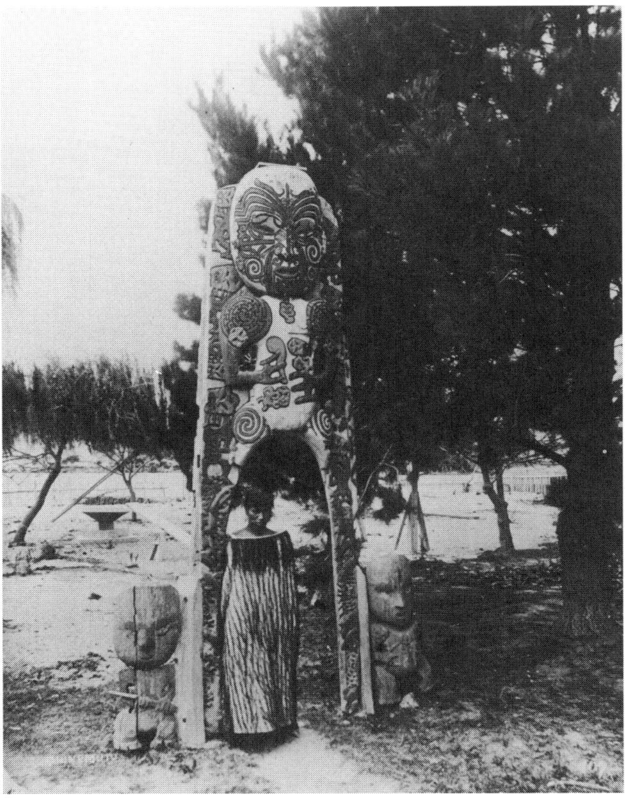

**Fig. 23.** The Pukeroa gateway of the Maori of New Zealand. Constructed for Pukeroa pa, Ohinemutu, in 1836, the gateway depicts Tutanekai, an ancestor. In Arawa moko, the father's side is the right-hand side of the face and the mother's is on the left. Courtesy of the Auckland War Museum, Auckland.

(Marguerite Duras 1984). The portraits, however, also had to record the deceased's face realistically if they were to function as ritual objects (Stuart and Rawski 2001). If anything in the portrait was not depicted accurately, then the ritual could be misdirected to someone else's ancestor. Because the person being painted had usually died before the painting was commissioned, this was not always an easy task.

The ancient Maya Codex Madrid depicts not only ancestors but also the burial of ancestors. Altar 5 at Tikal shows a skull and stack of four long bones. The underlying script describes a particular female, and the glyph band encircles what appears to be an exhumation scene, possibly for a second burial. Dates on the stela indicate that eight years passed between death, exhumation, and probably a second series of burial rituals. This information was important enough to be recorded.

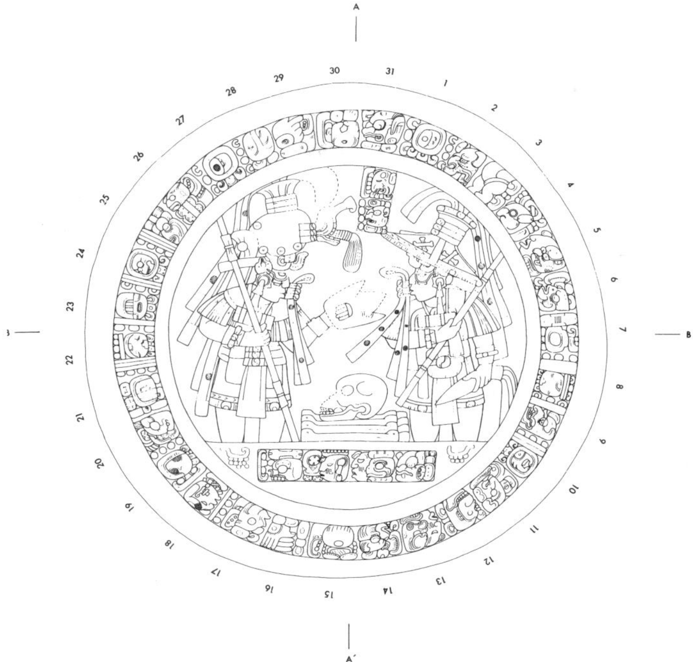

**Fig. 24.** Tikal, Altar 5, showing a skull and a stack of four long bones with an underlying glyph band describing a female elite personage. Courtesy of the University of Pennsylvania Museum.

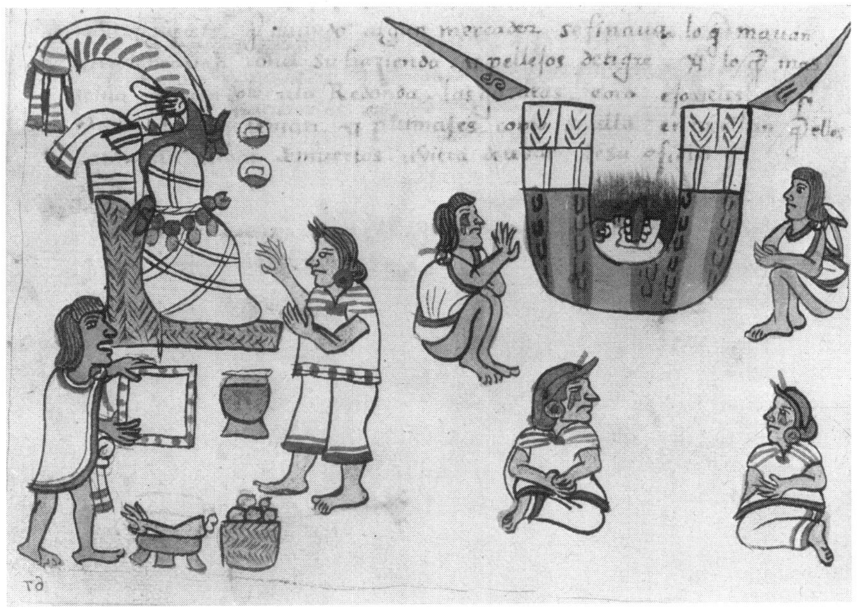

**Fig. 25.** Aztec funeral rites. Courtesy of Scala/Art Resource, New York.

The codices of the Aztec were considered sacred documents because they recorded genealogies, as well as the history, astronomical charts, divinatory tables, calendars, and religious ceremonies of their people. In a funeral ritual depicted in the Codex Magliabechiano, a man's children and relatives mourn his death, giving him *cacauatl* (chocolate) for the journey. On the side of the depiction, one can see the hole prepared for the burial.

## ART AND THE DEPICTION OF GENEALOGY

Visual art can be used to trace ancestry back to the identified common ancestor. In Rurutu, one of the Austral Islands, for example, there is a concern with genealogy and descent from ancestral deities. To record their genealogy, a wooden sculpture that has a masklike face is carved. Tiny human figures are placed in various positions on the body. The large figure is said to be that of the original ancestor of his descendants; the tiny figures represent those ancestors born after his death. The back of the sculpture is detachable, and originally the hollow body was filled with small figures of the humans who were descendants of the first ancestor (Barrow 1979). Among the ancient Maya, the dynasty of Tikal boasted a genealogy of thirty-nine sequential rulers who descended from the quasi-legendary lineage founder Yax-Moch-Xoc. These rulers' lives were recorded in hieroglyph texts. In Andean South America, ancestral figures were called *huacas,* which means sacred. People traced their lineage back to certain *huacas;* a shared *huaca* indicated shared ancestry (see Arriaza 1995).

## The Sacredness of Ancestress Objects

*Sacred* is often the word used to refer to anything associated with ancestors. The *mardayin miny'tji* of the Yolngu of Australia are sacred paintings in which ancestral beings are represented in figurative form. These designs, the Yolngu claimed, originated in ancestral times and depicted actual ancestral events (Morphy 1993, 247–48). This association with the ancestors, they claim, makes the paintings sacred.

Anything associated with the Australian totemic ancestors, a shield or boomerang or stone knife, is spoken of as being *churingas*—that is, sacred objects. *Churingas* used in rituals are "ornamented with a design that has a very definite meaning" (Spencer and Gillen 1912, 210). The only way to decipher or learn the meaning of any individual *churinga* was to ask a man of the totem to which it belonged. Stories about *churingas* were acquired during the initiation process; only initiated men of the totemic group know the meaning of the designs.

*Churingas* were surrounded by mystery. No woman, child, or uninitiated boy was allowed to handle or even see the sacred objects. *Churingas* of each local group were stored in a secret place, such as a cleft in the rocks that bound the sides of a deep gorge. It is of interest that the sacred Hopi ritual objects are also stored in a similar manner. In both cases, only the old men know the exact place where these sacred items are stored. For the Aborigines, this secret place was called the *ertnatalunga,* which was a house of refuge. Everything in the immediate neighborhood of the *ertnatalunga* was sacred and could not be touched or damaged in any way. The *churingas* stored in *ertnatulungas* are said to have belonged to the old ancestors who sent them into the earth at that special place. These *churingas* were carefully watched over, and every now and then they were taken out, examined, and rubbed with grease and red ochre. Many were so old that the designs were completely obliterated by touching and rubbing (Spencer and Gillen 1912).

It seems obvious from this discussion that it is difficult, if not impossible, to study traditional, prehistoric, or ethnographic art without studying ancestors. Despite their importance, ancestors have been the focus of only a few select studies. There is no theoretical approach that would lead us to argue that ancestors might be important. Social organization based on ancestors and descent is seen as primordial, despite the fact that we no longer honor Louis Henry Morgan's ides of cultural evolution beginning with such organization. Sacrifices for ancestors, given our assumptions of self-interest, make little sense, economic or otherwise. Yet we are left with the fact that ancestors and art are often associated and that this practice, which is linked to ancestral religion, is widespread. Steadman, Palmer, and Tilly propose, in fact, that ancestor worship is a universal tradition (1996).

The process of acquiring a technique was informal and involved a long period of time. During this time, social relationships were built, and other traditions, those related to social living, were acquired. The means—the process of or strategies for transmitting visual art techniques—are as important as the end, the work of art. The origin of visual arts, such as painting, sculpture, and architectural adornment, may owe a great deal to rituals associated with the dead. Masks may have developed out of the practice of decorating skulls. Sculpture may have developed out of attempts to depict ancestors. House ornaments may have been copies of the art used for tomb decoration.

The evidence presented thus far in regard to visual art indicates that to the extent decoration is inherited, it necessarily communicates ancestry. Further, ancestry and descent, kinship, religion in the form of ancestor veneration, and visual art style are all closely linked in the ethnographic and traditional visual art records. There also is a link between traditional visual art and moral systems that encourage kinship behavior among codescendants and that define the characteristics of a hierarchy, including the obligations and duties that define the role of the one at the top as well as the ones beneath. As visual art comes from the ancestors, influence is placed in the hands of the elders, who have mastered techniques. The transmission of knowledge places obligations not only on the elders but also on the young. Traditions require the formation and maintenance of enduring social relationship between the more conservative elders and risk-taking youth. Visual art traditions, in sum, require and promote social relationships among kin and metaphorical kin.

# CHAPTER 3

# Changing Styles of Visual Art

I'm sure that when people like Seurat started to do something, they really just wiped the past right out. Even the Fauves, even the Cubists did it.

—Marcel Duchamp

I follow many other scholars in arguing that when art history is viewed across the ages, what is obvious is continuity and resistance to change. What is unique in art history is the explosion of creativity and the evidence of change seen in Greek and Roman art and in much of the visual art produced since the Renaissance. Ancestors are the key to persistence; loss of ancestors, the key to change.

Change in visual art style that occurs when ancestors remain important, can be related to random errors in replication. Such changes will occur slowly and are likely to occur when codescendants come to be geographically isolated. When these changes are viewed in the largely discontinuous archaeological record, they can be seen as more profound and rapid than they actually were. In Ecuador's coastal rainforest, the Chachi make baskets that resemble, but are not identical to, baskets made by the Awa and the Tsachila. The Awa, Tsachila, and Chachi share a common but distant ancestor. Similarities here are related to common descent; peculiarities presumably are related to geographic separation, changes in available resources, and accumulation of random errors of replication (Coe 1995).

If distinct cultures, with distinct sets of ancestors, come together, visual art style seems to follow one of two basic trajectories. In the first trajectory, change will occur when strangers become kinsmen and metaphorical ancestors are created. In such cases, visual art will communicate this new ancestry. This has occurred many times in many places (for example, the Inca and their creator god; the Plains Indians and the calumet ceremony), and the prophets who created kinsmen of members of disparate tribes with distinct ancestral gods by uniting them under a more powerful creator god. Indian painters in the New World, living in places such as Quito, painted Christian religious figures and gave them Indian features. The artists, in so doing, depicted metaphorical ancestry. The visual art was used to encourage kinship behavior, such as honoring the new ancestors and being generous to and sacrificing for kin, including new kin. While often used to promote cooperation between metaphorical kin, visual art also can be, and often has been, used to encourage animosity against nonkin.

In the second trajectory, change in style also occurs when, due to contact with people with distinct cultural practices, traditions are discarded, along with the ancestors who encouraged those traditions. When there are no ancestors said to have expectations about how descendants should behave, the arts proliferate and become more elaborate. In other words, when traditions no longer restrain competitive behavior, the door opens to competition. As it is apparently less threatening to change a technique than it is a theme or motif, change begins when artists emphasize and competitively develop techniques. If there is a reward for technical competence (it attracts the attention of a wealthy collector), not only will new techniques proliferate, but new styles, such as still life and landscape paintings, also will begin to be developed. Evidence of increasing competition consists of a runaway type selection; the visual arts move toward increasing elaboration or outrageousness. Baroque and rococo styles are examples of such exaggeration.

When an art style changes rapidly, we are likely to find a series of countermovements, trying to move visual art and related behaviors back to ancestrally sanctioned ones. Tutankhamun restored the old religion and art of Egypt replacing that introduced by the heretic pharaoh Akhenaten. Savaronola attempted to motivate the artists of Renaissance Florence to return to themes of the past. Church leaders during the Council of Trent encouraged a return to the simple, religious art such as that encouraged by Pope Gregory the Great. This has happened in more recent centuries, when social movements tried to erase art considered decadent.

Visual art produced during four periods supports these descriptive claims: the art of the Greeks and Romans; the art of the Middle Ages; the art of the Renaissance, with its transition from Christian art to a European style; and art produced since the beginning of the twentieth century, with the transition to contemporary art. In studying change, I focus primarily on the West because the tremendous persistence over thousands of years that has been observed in the East has not been known in the West (Gombrich 1989, 125). The West has experienced a series of reactions against the past that began with the Greeks and has continued, with occasional interruptions, to the present day.

The Greeks, who initiated many social and cultural changes, worked to dilute the power of the old clans and clan ancestors to establish a new social order. While this experience was not completely unique (earlier societies, such as the Sumarians, had initiated cultural changes), new forms, motifs, and styles of visual art were developed, as were new techniques. Greece went on to influence not only the Romans but also much of Europe and, through Europe, much of the rest of the world.

During the Middle Ages the artists of Europe turned to new ancestors, religious ones this time (God the Father, Mary the Mother, and Christ the

Son). In Christian art, these metaphorical ancestors not only provided models for behavior but also created metaphorical kinsmen and kinswomen out of the people of Europe. For almost a thousand years, religious teaching about proper kinship behavior was a primary aim of visual art.

The Renaissance, on the other hand, "was a time of warfare between the old and the new. . . . Almost every Renaissance writer, humanist, artist, and thinker was in a sense a rebel and to some extent an iconoclast." The rebirth implied by the word *Renaissance* was a rebirth of the secular, of the culture of Greece and Rome (Ergang 1967, 9, 11). Despite the fact that universities had their origin during the Middle Ages and Petrarch (1304–1374) and Giotto (1267–1337) are credited with inspiring the Italian Renaissance, everything medieval was now mocked. The term *medieval* became an adjective that described anything obsolete or unprogressive. *Dunce* was a word that developed out of derogatory comments about Duns Scotus, an outstanding scholar of the Middle Ages. Writers of the Middle Ages now were dismissed as "obscure" and "inane" (8). Renaissance artists, free of the traditions that restrained their creativity, were seen as guides of the new social order.

## Greece: Loss of Traditions and Ancestors and Proliferation of the Arts

Greece initially was a tribal society made up of distinct people linked to one another by the relationships of city-states that shared a language. Over time, these tribal units, founded by particular ancestors, began to dissolve along with the traditions that had encouraged tribal cooperation. The Greeks turned away from the "primeval clan organization." Life in a Greek democracy had to become "free of all rigid traditions" (Hauser 1959, 58, 89).

It was in Athens around the sixth century B.C.E. that the revolution, Greece began to change rapidly due to a political and economic resurgence that resulted in an influx of craftspeople (Lucie-Smith 1992, 63). The changes initiated at this time would lead to "the greatest and most astonishing revolution in the whole history of art." Prior to the sixth century, "artists of the old Oriental empires had striven for a particular kind of perfection. They had tried to emulate the art of their forefathers as faithfully as possible, and to adhere strictly to the sacred rules they had learned" (Gombrich 1989, 48). By the sixth century B.C.E., this was no longer true.

During the classical period (480–335 B.C.E.), artists, who increasingly were unwilling to follow a formula, began to experiment. Without the traditional formula, there was a rapid stylistic evolution that placed an emphasis

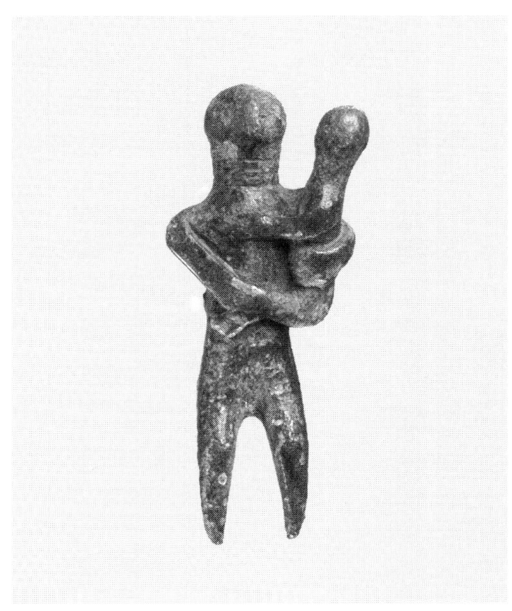

**Fig. 26.** Mother and child (bronze), allegedly from the Cyclades, 1150–1000 B.C.E. Courtesy of the George Ortiz Collection, Switzerland.

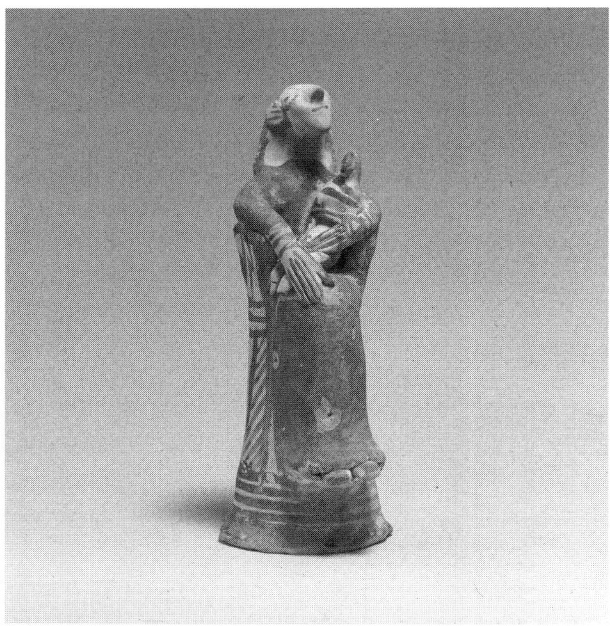

**Fig. 27.** Mother and child (terra-cotta), Northern Attic or Boeotian, circa 700 B.C.E. Courtesy of the George Ortiz Collection, Switzerland.

on technical competence. Sculptors "tried out new ideas and new ways of presenting the human figure, and each innovation was eagerly taken up by others who added their own discoveries" (49). Individual artists, some of whom now signed their work, emerged and came to be celebrated. The public became fascinated by art done for its own sake (to show the technical competence of the artist) and not done to serve religious or political functions (65). This emphasis on technical competence and experimentation appeared in the other arts, including architecture. With this innovation, artists had "set out on a road on which there was no turning back" (49).

Classical dramatic tragedies show the conflict between loyalties to the clan organization, the ancestors, and loyalties to the state. Sophocles' (496–406 B.C.E.) story of Antigone revolves around the problem of loyalty to ancestors and obedience to new laws aimed at destroying such loyalties. The story unfolds when Polynices, Antigone's brother, is killed in his attack on Thebes and must be buried. While the new laws prohibit such action, Antigone, as his female kin, is bound by ancestral law to do the burying. When Creon's men discover Antigone in the act of burying her brother, they capture her and she is buried in a rock chamber, where she dies.

Ismene, Antigone's "true sister-in-descent," had let her fears keep her from helping her sister and fulfilling ancestral obligations (Fox 1993, 160). Yet, when Antigone died, Ismene tearfully begged Creon to bury her alive with her sister. "This act must have drawn a gasp of relief, admiration and approval from the seventeen thousand Athenians in the Theater of Dionysis" (151). What this audience knew, and many today may not, is that these sisters, as the "last living members of the line of common descent have all the duties of the women of the line thrust upon them. It is their burden and they must share it totally and without reservation" (159). Antigone and Ismene were descendants of the house of Cadmus, the last living children of the cursed line through Oedipus (146). The line was cursed because Laius, the father of Oedipus, destroyed his own family due to his own selfishness in carrying off Pelops' young son Chryssipus, who later hanged himself in shame. The acts of the father, in other words, can bring tragedy to their children.

Socrates (470–399 B.C.E.), rather than admiring traditions, questioned them along with the authorities who encouraged them. Rationality, he argued, was the way to understand the universe. Influenced by Socrates, Plato (428–347 B.C.E.) wrote the *Republic*, a book best known for its political theory, not, as Miller (1999) claims, its abstract philosophical truths. The *Republic* is, Havelock (1963) argues, political in the sense that it is "the first systematic ordered educational curriculum proposed in the west" (Brown 1972, 673). The plan outlined in *Republic* was aimed at producing a sound body and sound mind. This involved, among other things, destroying traditions and encouraging rational thinking and skepticism.

While Plato's discussion focuses on the art of poetry, it may be applicable to all of the arts. Plato rejected Homer's epic poems *The Iliad* and *The Odyssey*, arguing that they are educational texts that were "too powerful in educating the young" (Brown 1972, 672). Students who heard these poems skillfully and socially recited were inspired not to analyze, consider alternatives, or ask questions such as "What is truth?" or "What is justice?" Instead, they were inspired to imitate or to copy (mimesis). Homer's heroes, the metaphorical Greek ancestors, became personal role models for students; each time an epic was recited, these models of behavior were revalidated. Homer's heroes, Plato argued, influenced youth into "engaging in undesirable behaviors" or behaviors governed by passion, not reason (673).

Plato's argument raises several issues. First, he argued that art (in this case poetry) could have a powerful influence on behavior. Second, he proposed that if we are to establish a new order, it is necessary to challenge ancestral ways that encourage particular loyalties and social behavior. Once ancestors and traditions are devalued, individualism, creativity, and the arts begin to flourish. Third, Plato and Socrates recognized that artists freed of traditional social constraints made art that popularized the artist, even if that art, by feeding the passions and impairing reason, influenced behavior that was at the expense of social order.

Plato's argument, convincing as it may sound, apparently had little influence on artists. The Hellenistic period—which is said to extend from the accession of Alexander the Great to the throne of Macedon in 336 B.C.E. to the death of Cleopatra VII of Egypt in 30 B.C.E.—saw the development of innovative and often exaggerated art. During the Hellenistic period, technique and style were further elaborated. Still lives and landscape paintings were produced and wealthy Greeks paid fabulous prices to collect works of art (Gombrich 1989, 77). Praxiteles became known for the beauty of his sculpture; Scopas designed the Great Alter of Pergamum (ca. 180 B.C.E.), which depicts a battle between man and the gods and giants. Due to its theatricality (or exaggeration), this sculpture has been described as baroque, a term used to refer to the exaggerated art of seventeenth-century Italy (Lucie-Smith 1992, 61).

Greek artists had a significant influence on the art of Rome, which scholars often describe as classical art—that is, as heir to Greek art. Roman art was "Greek in many of its most important manifestations," although it was influenced by Roman contact with other cultures (Gombrich 1989, 72). The Romans recognized the power of art as propaganda, using it to proclaim victories in war and to intimidate.

## The Middle Ages

The Middle Ages began, some argue, when barbarians, mainly nomadic Germanic tribes, invaded the reaches of the Roman Empire and threatened Rome's northern borders. While the Roman emperor only had subordinates, the barbarian chief was surrounded by a band of companions *(comitatus)* who were bound to him by ties of kinship, or blood, and who considered themselves to be almost his equal. The Germanic tribes, once settled in the new lands, adopted the local religion, often Christianity, and dismissed the Romans as effete and servile. The traditional tribal structure of the Germanic tribes was elaborate, evolving into a feudal system based on the kinshiplike bonds a chief had with his companions (Lucie-Smith 1992, 107).

Jewish law, which influenced Christian law, forbade the making of images, although Jewish colonies, such as the one at Dura Europos, were known to decorate the walls of synagogues with paintings of sacred stories (Gombrich 1989, 89). The early Christians placed paintings in the catacombs of Rome (around the third century A.D.), which were Christian places of burial. Christian leaders, however, did not completely trust visual art; statues were too likely to resemble the graven images and heathen idols condemned in the Bible. In the sixth century, however, Pope Gregory the Great changed the direction of art by proposing that "painting can do for the illiterate what writing does for those who can read" (Gombrich 1989, 95). Art was to be used to educate. Art, however, now had to be simple; nothing could be placed in a painting that was not central to the theme or story. This sacred story was held to be more important than the artist's creativity or technical competence. Even though artists must have been familiar with Greek art, their work now became simple; figures, for example, often began to look stiff and rigid (96).

While the art produced under the influence of the papacy was now used as a conspicuous reminder of the teachings of the church, the visual arts of Byzantium followed a different path. As a consequence of iconoclasm (A.D. 711–843), a political and religious movement hostile to all sacred images, all religious art was forbidden by the Eastern Church by the year 745 (Lucie-Smith 1992, 92). Eventually, opponents of iconoclasm began to emerge, arguing that images were not only useful for teaching but sacred. Visual art was seen as a "mysterious reflection of the supernatural world." The thinking underlying this claim was that "If God in His mercy could decide to reveal Himself to mortal eyes in the human nature of Christ, why would He not also be willing to manifest Himself in visible images" (Gombrich 1989: 98). As visual art was a reflection of the sacred, artists could not follow their fancy. The true sacred image, or icon, was one hallowed by age-old tradition.

By the early 1300s, the church was at its zenith, visual art was its handmaiden. "Medieval painting and Christian painting were one and the same" (Francastel 1967, 7). The visual art of Middle Ages in Europe, however, was not without innovation. Gothic architecture was introduced, and during this period Giotto di Bondone emerged as a well-known painter and architect. Giotto often is regarded as the founder of Western painting because his work broke free technically from medieval restraint and the stylizations of Byzantine art. In 1372, Cennino Cennini (a pupil of Giotto) wrote one of the first books on art; predictably, it was a treatise on techniques.

Scholars such as Petrarch reintroduced classical ideas, and these ideas, when combined with an increasing emphasis placed on technique, began to influence the changes that produced the Renaissance. Although the majority of the art of the Middle Ages was religious, not all of it existed exclusively to serve the church. One of the few remaining secular works of art, the Bayeux tapestry (ca. 1080), is about kinship and ancestry. This tapestry, while not religious, tells the story of the Norman Conquest. In one section, Harold swears an oath of fealty to William by placing his hands on ancient and sacred ancestral relics. Across Europe, family members placed stone carvings of knights on their tombs. While much of this period's art was about kinship and ancestry, statues began to be carved of highly placed persons, such as Can Grande della Scala, (1291–1328), the Imperial Vicar, Lord of Verona, and an early art patron.

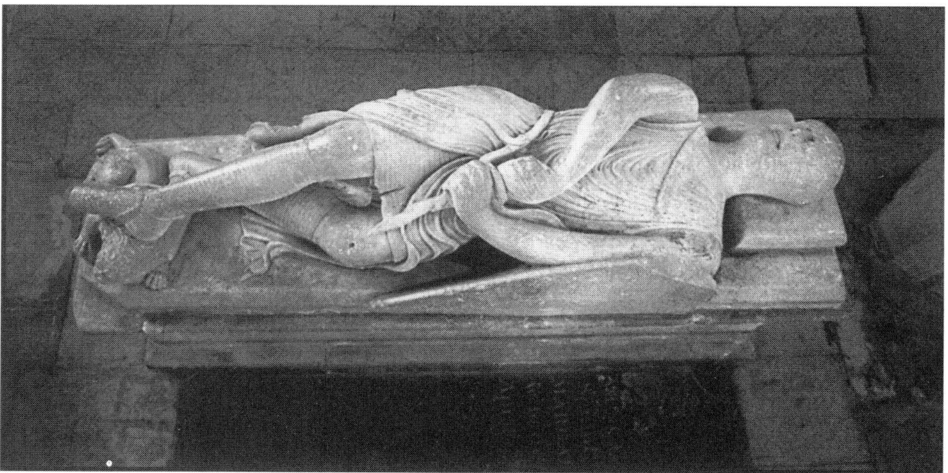

**Fig. 28.** Medieval effigy of a knight (Sir John Holcombe?), circa 1280, in Dorchester Abbey, Oxfordshire. Photograph by Frank Blackwell.

## The Fifteenth Century

The beginning of the fifteenth century was a transitional period, one in which the ideas that would shape the Renaissance were percolating, especially among the generation of artists active in Florence. New ideas, especially those reborn from the Greeks and Romans, began to influence new forms of secular art and to influence religious paintings. With the invention of printing around 1450, millions of copies of the illustrated Books of Hours, which were basically religious texts, were printed (Febre and Martin 1976). "Devout or not, all owned and carried Books of Hours" (Tuchman 1978, 236).

Books not only helped with devotions; they helped revive classical ideas of logic and science. Of the books printed between 1450 and 1500, 75 percent were ancestral in the sense that they were religious (45 percent) or dealt with classical and medieval historical and literary topics (30 percent). Twenty percent addressed law and scientific subjects such as medicine (see Febre and Martin 1976).

In 1435, Alberti published his widely circulated treatise on painting *(Della Pittura)* to try and move painters still using the traditional methods of the Middle Ages into the principles of the new art. Knowledge of geometry, Alberti argued, was essential for artists. Alberti described the artist as a learned gentleman, a solitary genius (Chadwick 1990, 15). These genius artists needed to be freed from "gild restrictions" that influenced ordinary craftsmen; they needed to be independent so that their genius could flourish (Ergang 1967). "One must treat these people [artists of extraordinary genius]," Cosimo de Medici poetically explained, "as if they are celestial beings, and not beasts of burden" (Ergang 1967, 90).

In 1438, Plato was translated into Latin. Cosimo de Medici was so intrigued by Plato's ideas that he founded the Platonic Academy in Florence. This academy was secular and had as one of its roles the challenging of traditions. Wealthy beyond imagination, the Medicis were the source of avant-garde thinking, commissioning artists who painted their Neoplatonic view of the world.

Botticelli studied under Fra Filippo, one of the last devotional, or religious, painters. By 1470, Botticelli, however, had embraced Neoplatonism, and the Medicis felt that he could be their artist representative. He was brought into Lorenzo's court and commissioned to paint Saint Sebastian for the central nave of the Church of Santa Maria Maggiori. Soon, however, Botticelli became one of the first artists to "enlarge the sphere of art by choosing purely secular subjects from classical mythology" (Ergang 1967, 67). He painted *Primavera* (which depicts a Madonnalike Venus in a bacchanalian scene) for the Medici villa at Castello.

Botticelli soon devoted himself to painting allegories that celebrated the Medicis. In *The Adoration of the Magi,* he painted the likeness of a man believed to be Lorenzo the Magnificent. His *Young Man with a Medal* is considered to be a tribute to the Medici, because the model holding the gilded medal of Cosimo de Medici is believed to be the philosopher Giovanni Pico della Mirandola, who, at the encouragement of the Medici, published a series of 900 questions challenging the church. Soon thereafter, other artists began painting images of the Medici in religious scenes.

The art patrons, rather than church elders, now began to determine the content of paintings. Not only were patrons depicted in paintings, but the patron's interests also influenced the theme of paintings. The primary duty of man during the Middle Ages was to serve God and his or her neighbors, who were brothers and sisters in Christ. The first duty of man during the Renaissance was to himself and to his improvement or self-perfection; morality thus was divorced from religion and art (Ady 1993). Nudity was now seen as a virtue. No longer were a naked Adam and Eve ashamed of their nakedness. Boticelli's Venus, as depicted in his painting *The Birth of Venus* (ca. 1485), stands proud of the perfection of her body. The human body

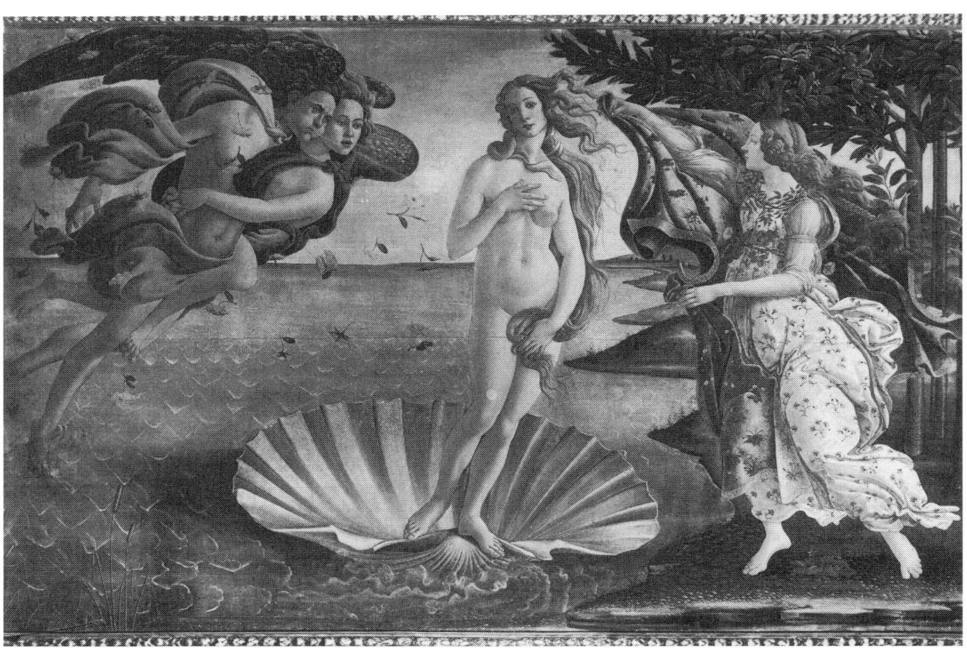

**Fig. 29.** Sandro Botticelli (1444–1510). Courtesy of Uffizi the Museo di Andrea del Castagno, Florence, Italy, and Alinari/Art Resource, New York.

during the Renaissance was, as it had been with the Greeks, "a proper means of representing perfection" (Hollander 1978, 96).

It is perhaps ironic, as Botticelli played an important role in changing art during the Renaissance, that by the High Renaissance, his work was considered passé. Despite the voluptuousness of his nudes and the bacchanalian scenes he painted, Botticelli was dismissed as a moralist. Botticelli had come under the influence of Savaronola, known for his denunciations of the secularism of life and culture. Under Savonarola's influence, Botticelli was led to denounce the secular joys of the flesh seen in *Primavera* and *The Birth of Venus*. Crippled and forgotten, he died in 1510. His work was neglected until John Ruskin and the English pre-Raphaelites rediscovered it (Gramercy Great Masters 1996).

## The Sixteenth and Seventeenth Centuries

The High Renaissance in Italy, which lasted only from about 1495 to 1520, produced the greatest number of notable artists, among them Leonardo da Vinci, Michelangelo, Raphael, and Titian. Prior to this period, technique was so important that Leonardo da Vinci is said to have used a mirror to assess his drawings' accuracy (McEvilley 1992, 101). Discussions of technique, however, soon began to be replaced by discussions emphasizing new themes. There was a new sense of art as being not a copy of nature but an ideal construction. Visual art was regarded as an improvement on nature-governed by its own laws, which were separate from the laws of nature (Goldwater and Treves 1945, 13).

As less emphasis was placed on technical competence, and more was placed on what made a painting attractive to the viewer, the idea that artistic ability was determined not by schooling or practice but by a special endowment began to spread across Europe. Technical rules and formula now were of little importance compared with an inner vision that drew upon "wonderful and divine thoughts" that combined ecstasy with the operation of the intellect (Vasari 1984, 204). The production of a work of art now had to be driven by irrepressible urges, perhaps even a mixture of fury and madness akin to intoxication.

Artists were divine beings (Kris and Kruz 1979) who had a special, prophetic vision. As Durer (1471–1528) explained, "God is honored when it is seen that he has bestowed such genius upon one of his creatures [the painter] in whom such art dwells" (Goldwater and Treves 1945, 67). Social rules did not apply to genius; genius excused artists from normal discourse and social life. "Excellent painters," Michelangelo (1475–1564) argued, "are not unsociable from pride, but either because there are few minds capable of the art of painting or in order not to corrupt themselves with the vain conversation of idle persons and degrade their thoughts from the intense and lofty imagining in which they are continuously rapt" (Goldwater and Treves 1945, 67).

The public, while increasingly enamored by artists, books, and new ideas, was growing skeptical about ecclesiastical leaders, who were seen as unresponsive to the concerns of the people. In October 1517, Martin Luther denounced the sale of indulgences and the theological substructure upon which the practice was based, and he proposed to the hierarchy of the church his ninety-five theses, which were printed and widely disseminated. Clergymen, who often had the advantage of being able to read, began to see the church as too conservative and started pushing for reform. Altar paintings were commissioned from openly heretical artists. Venetian painters began to give "nominally religious themes a distinctly secular treatment" (Dewey 1958,188). Michelangelo, regarded as a religious artist, painted (for the Pontifical Chapel in Rome) the entire Holy Family, including the Virgin Mary, naked (Goldwater and Treves 1945, 108). With artistic freedom came the idea of a sexual relationship between the painter and his model. According to a novel written by Bandello (1554), the artist's sexual relationship with his model was the "artist's prerogative" (Kris and Kurz 1979, 117). By 1527, the sack of Rome by Charles V of the Holy Roman Empire ended the patronage system there, scattering artists and scholars in all directions (Ergang 1967; Chastel 1983).

The Council of Trent, held between 1545 and 1563, was initiated to fight the corruption in the Church and to place limits on visual art produced at the encouragement of members of the church hierarchy that implicitly or explicitly disparaged religion. The twenty-fifth session of the council (December 3–4, 1563) concluded that the arts had to be made virtuous. Visual art's role was not to entertain but to encourage appropriate social behavior through realism, or narrative paintings. These works of art provided behavioral models for viewers to imitate. Visual art should be used solely to instruct viewers about the Holy Family, the saints, and mysteries of redemption. Paintings depicting naked saints or members of the Holy Family involved in licentious poses were strongly discouraged; and artists were commissioned to paint gauzy drapery over the nude figures painted in earlier works.

While the council recommended ecclesiastical control of images (artistic freedom was to be limited and artists were to be strictly supervised by the church), their recommendations often held little power. Veronese was summoned before the tribunal of the Holy Office ten years after the Council of Trent ended (July 1573) and was charged with introducing disrespectful and fanciful details in one of his paintings of the Last Supper. He replied to their questions of why he had included these "scurrilities," saying, first, that the painting had "unfilled space" that bothered him, and, second, that "we painters take the same liberties as poets and madmen take" (Goldwater and Treves 1945, 106). He was asked to remove from his painting of the Last Supper a dog, a dwarf, a parrot, German halberds, and the "bleeding nose" of a servant. The way Veronese answered the tribunal's questions, and his general lack of concern in following their demands (he only painted out the bleeding nose), suggests not only that he, as many artists of the time, had high-placed protection but also that the council could safely be ignored.

During the seventeenth century, visual art continued along the path that emphasized genius and creativity. It was during this century that a widespread taste for elegance, comfort, and beautiful objects was developed. Louis XIV of France advocated childlike behavior characterized by freedom, lightheartedness, and irresponsibility. His court was highly artistic and frivolous. His taste for elegance, indulgence, and the new and his disdain for self-restraint infiltrated not only the nobility but also the ranks of the bourgeoisie.

## The Eighteenth Century

Louis XIV died in 1715. Partially due to the excesses of his court, salons emerged early in the century and provided a forum at which philosophers could publicly question authority (Goldwater and Treves 1946, 204). Eventually, due to widespread corruption in society, visual art began to take on a strong political focus; artists began to get involved in social criticism.

Jacques-Louis David, who was the voice for social reform in France (1748–1825), wrote that the arts must "contribute forcefully to the education of the public." For David, this meant that he "would help the arts towards their true destiny, which is to serve morality and elevate men's souls" (209). Hogarth (1697–1764) became well known in England for paintings that satirized life in the modern city and were sympathetic with professional women such as whores and fishmongers who faced many dangers. Goya (1746–1828) created the first of his print series *Los Caprichos* between 1797 and 1799. This series mocked the social mores and superstitions of his time and satirized the vices and follies of women, lawyers, physicians, and the church, priests, and monks in particular. He also pointed out the horrors of war.

During the eighteenth century, the term *ästhetik* was coined by the philosopher Alexander Baumgarten, and a separate area of philosophy emerged based on Plato's ideas of beauty (Dickie and Sclafani 1977, 2). Emotion became important in discussions of art; as a response to beauty, emotion was now seen as a crucial element of visual art. As Diderot, who

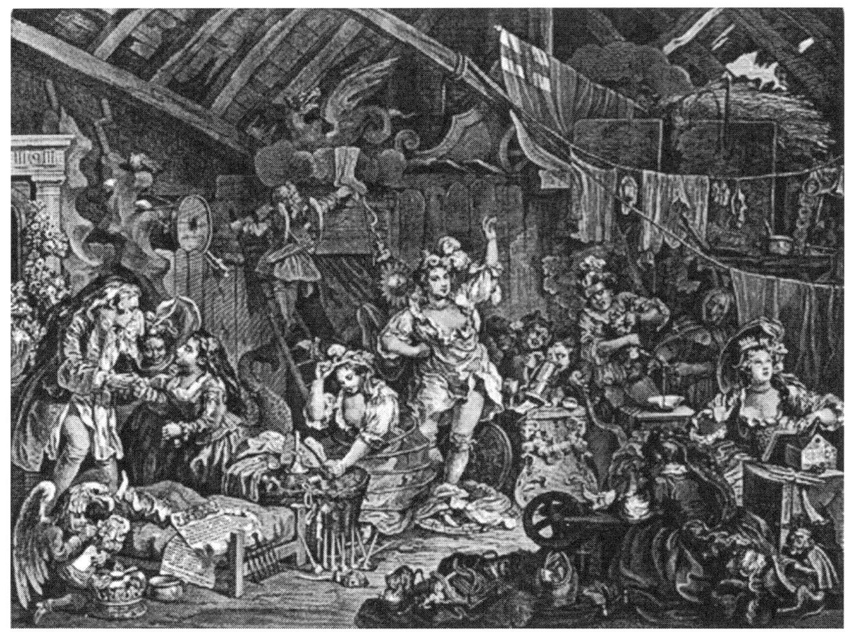

**Fig. 30.** *Strolling Actresses Dressing in a Barn,* circa 1738, by William Hogarth. Courtesy of Deering McCormick Library of Special Collections, Northwestern University, Evanston, Illinois.

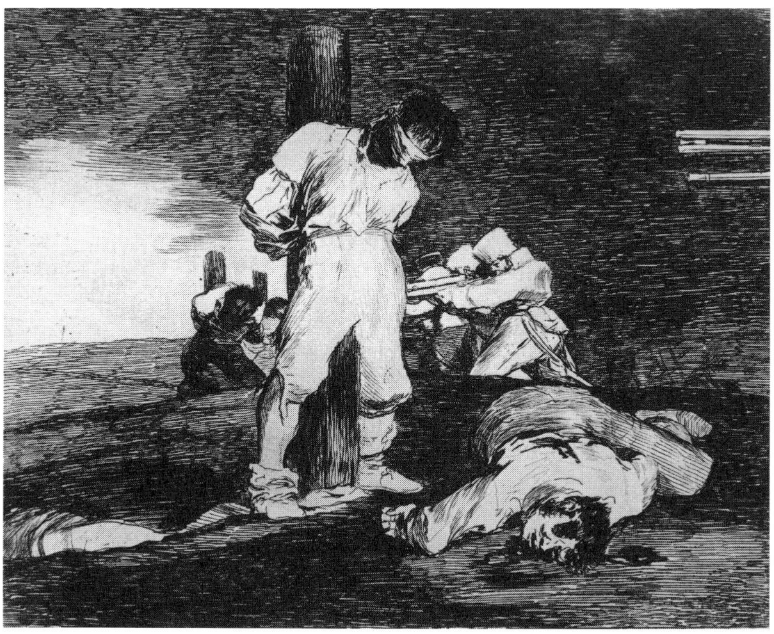

**Fig. 31.** *Disastre de la Guerra 15: "And it cannot be changed."* Francisco de Goya y Lucientes. Courtesy of Foto Marburg/Art Resource, New York.

reviewed the Salons, expressed, the "chief business of art is to touch and to move" (Holt 1958, 311).

## The Nineteenth and Twentieth Centuries

Influenced by the rising merchant class and a growing number of art collectors, art styles began to change, driven by collectors who were intrigued by novelty. When a new art style came into fashion, the "works of all artists directly or indirectly connected with it were almost equally sought . . . It was almost as if no critical sense existed" (Seligman 1952, 4). "Even more surprising is the sweeping manner in which all were discarded overnight . . . Taste [evolved] by a series of fanatical infatuations and complete rejections" (vii).

Art styles now lasted only a short period of time. A few decades after David painted *The Crowning of Bonaparte and His Empress,* Constable, viewing this painting, wrote: "As a painting it does not possess anything of the Language of the art, much less of the oratory of Rubens or Paolo Veronese; it is below notice as a work of execution" (Goldwater and Treves 1945, 268). Almost fifty years after Goya completed his *Disastres de la guerra* (Disasters of war), Edouard Manet dismissed Goya's work, calling it "inferior" (303). Innes, in a letter written to the editor of an unknown newspaper, referred to impressionism as a "humbug," a mere "pancake of color" (344).

While artistic innovation was encouraged, this encouragement was not extended to narrative paintings (that depict a story and emphasize subject matter) and naturalistic representations (realism), these were disparaged, partially as a reaction to the earlier use of art as propaganda to promote social movements and war. Replacing narrative was an emphasis placed on color, line, and form, along with the emotions they inspired. Eugène Delacroix became famous primarily for "using a color palette capable of eliciting specific emotional reactions from his viewers" (Fleming 1974, 316).

Art styles soon changed so rapidly that the public lost confidence in its ability to judge art; it was no longer clear what distinguished good art from bad. Critics stepped into this void and began to use their influence to guide contemporary art and taste. In the 1850s, painters doggedly painted according to Ruskin's precepts. At the turn of the century, collectors bought art only from artists endorsed by Alfred Stieglitz (Maas 1984, 14).

An unrelenting search for something new in visual art characterized the nineteenth and twentieth centuries. Yet, as early as 1821, Ingres had issued a cautionary note: "Let me hear no more of that absurd maxim: 'We need the new, we need to follow our century, everything changes, everything is changed.' Sophistry—all of that! Does nature change, do the light and air change, have the passions of the human heart changed since the time of

Homer: 'We must follow our century': but suppose my century is wrong" (Goldwater and Treves 1945, 218). Although few people seemed to listen to or at least respond to this idea, it was repeated over time. In 1992, McEvilley argued that these endless changes in visual art were mutually trivializing: "If a great artistic movement is going to last a few years or months," he asked, "what really is the value of it?" (127).

Increasingly, greater emphasis was placed on the emotional response to art. Berenson argued that visual art should "tune us like instruments—instruments for ecstasy" by stirring "all of our emotions" (1948, 147). Ecstasy was aroused, he argued, not by narrative but by visual art's "significant form"—the combination of lines, forms, and colors that characterize visual art (Bell 1958, 40). Recognizing that we do not know precisely why line, color, form could "make people happy, sad, excited, and so on" (Blocker 1979, 101), Langer proposed that the response actually is to the symbols (1953, 40). "Art," she argued, "is the creation of forms that are symbolic of human feeling."

Avant-garde art and modern art, to a large degree, were built on the destruction of the visual art that had been created earlier. Men like Alfred Stiegliz, Clement Greenberg, and Hermann Broch dismissed traditional art, even paintings that previously had been regarded as great works. The younger generations of twentieth-century artists, Seligman explained, rejected all the work of the prior generations (1952, 57). The fact that money was clearly influencing artists who wished to be popular led one writer to express the following in *The Art World:* "The field of art is no longer an innocent, Sacred Elysium in which each artist, critic, and aesthetician labors for the greater glory of God and the salvation of mankind . . . The vast increase in millionaires . . . [has] opened up to the unscrupulous so many avenues of gain that it has transformed a once sacred field of ideal and poetic aspiration into a sordid mart of financial speculation. The money-changers have invaded and are desecrating the Temple!" (Anonymous 1918, 253).

Alfred Stieglitz, who played a seminal role in the development of modern art in America, established the experimental art gallery "291" in New York City. Open from 1908 to 1917, this gallery exhibited the work of artists such as Rodin, Matisse, Cézanne, Brancusi, and Kandinsky. In 1914, he mounted the first exhibition of African sculpture presented as art, not as material culture (National Gallery of Art 2000). Committed to American artists, Stieglitz devoted much of his time after 1913 to their behalf. In 1917, he bravely exhibited Marcel Duchamp's *Fountain,* which had been rejected by another exhibition (National Gallery of Art 2000). Due to the efforts of Stieglitz, artists again were regarded as seers, and the gallery was referred to as a church consecrated to those who had lost old gods and who had need of new ones. Artists as prophets (Lippman 1929) could lead us to a brave new world that was above commercial considerations by producing visual art that, Bell explained, "might prove the world's salvation" (1979, 47). The Dadaists, Hoebel (1949, 162) wrote, tried to do this by pointing out that "modern civilization [was] so false and meaningless that the honest artist can only lampoon and destroy it with senseless combinations of line and color."

> The artist far more than the layman is perceptive of his environment and this new potential has allowed to him a new vision of which contemporary art is a materialization . . . Genius' is a *rara avis* whose meteoric appearance revolutionizes the artistic world and, often unbeknownst to himself, adds a new and brilliant star to the aesthetic constellation. He has a message with which to astound the world and for which the technical means at his command may be too limited . . . Alone in his ivory tower of thoughts, alone with his torments, unconscious of the outer world's appreciation or deprecation, he follows his divine mission. (Seligman 1952, 153, 160)

The idea of the artist as a madman, hinted at by Veronese after the Council of Trent, was fine-tuned around the time of Freud; mental illness became linked with the creative process. Some felt that there was a causal connection between the psychic illness of the artist and his artistic power. Saul Rosenzweig refers to this madness as the "sacrificial roots" of art. Artistic power was gained through mental suffering (Trilling 1971). What was not clear was whether the madness of the artist was actual or a ploy to attract attention to his or her art.

Narrative paintings continued to be dismissed as lesser forms of art because they were seen as tools used by social and political groups to "rouse us to action" (Berenson 1948, 68). "From Plato's day to the latest," Berenson complained, "states, societies, synagogues, churches, and conventicles have deliberately tried to harness art for their particular advantage and, failing that, suppress it altogether" (147). Berenson did not deny that visual art could influence behavior, he just did not want to see it used that way. In fact, he wrote that narrative paintings "cannot help exercising influence, seeing how prone we are willynilly, to imitate what we see and be affected by what we hear" (22).

To summarize this brief discussion on visual art, the Greeks introduced the ideas of destroying traditions and encouraging skepticism and creativity. They influenced the Romans, who spread their ideas across their empire. The Middle Ages involved, among other things, a reaction against the seeming excess found in Roman art and society and a return to the association of visual art with kinship, ancestry, and religion. Greek ideas were reintroduced during the Renaissance and, with a few interruptions, artists since have attempted to free themselves from ancestral bonds, often referring to their attempt by using such terms as *self-expression*. New styles that attracted public acclaim resulted from their search. In order to attract attention, artists have had to be innovative.

## Traditional Versus Nontraditional Visual Art

Traditional visual art refers to art that, although perhaps technologically complex and attractive (meaning it attracts our attention), shows evidence of repetition and persistence across generations. Nontraditional visual art, on the other hand, refers to art that clearly or explicitly ignores or rejects the traditions of one's ancestors. Artists, whether they live today or in the past, who create idiosyncratic visual art are nontraditional artists. When French artists borrowed heavily from the African tribal arts to make a new style of art, the French painters, most would agree, were not making traditional art. The highly innovative and idiosyncratic mainstream visual art of the twentieth century, with all its isms (for example, Dadaism, impressionism, abstract expressionism), is nontraditional art.

Because traditional art copies the past, it is ancestral. While ancestry is important to many people around the world, beginning as early as the Greeks, ancestors have been discarded. Underlying history are stories of the rejection of ancestors; creation of new, metaphorical ancestors; wars between people with distinct ancestors; and collaboration among people sharing common ancestry, whether actual or metaphorical. While ancestry is not important in much of the West, ethnic wars, guided by symbols of ancestry, continue to affect our world and even our daily lives.

Admittedly, this dichotomy is not perfect. The visual artists who between the sixteenth and twentieth centuries explicitly rejected the art style (techniques, features) of their European ancestors continued to use some materials (for example, pigments, paints, and brushes) that came from the past. They could not completely escape traditions. Further, copying (meaning replicating or using the style of) is not unknown in nontraditional art. The terms *genre, school of art* (as in the school of abstract expressionism), *style,* and *artistic period* suggest that one artist has strongly influenced another. Picasso wrote, "I have no hesitation, when I am shown . . . a portfolio of old drawings, in taking from them whatever I want" (Zervos, 1970, 115). Jean-Auguste-Dominique Ingres asked, "who is there among the great men, who has not imitated. Nothing is made with nothing" (Goldwater and Treves 1945, 217). William Blake, the mystic, seer, poet, and painter, wrote that the "difference between a bad Artist and a Good One Is: The Bad Artist Seems to copy a Great deal. The Good One Really does Copy a Great deal" (263).

I assume that terms such as *genre* mean that artists are influenced by and copy other artists, even if, as is quite likely, they "seldom acknowledge that another work of art in the same medium" ever influenced their own work (Munro 1949, 347). While similarity in nontraditional visual art implies copying peers (or in the case of neoclassical art, copying the classical artists), in the traditional arts it means copying one's own ancestors.

Further, traditional art, while conservative, is not stagnant. The visual arts of China, for example, are said to have "a single, uninterrupted history," which apparently means that ancient styles and principles were regularly renewed or revived (Fong 1984, 2–7). Techniques, colors, forms, and motifs of ancestral art were a permanent referent in regard to appropriate style; any new elements had to complement the ancestral ways. While new elements are not strictly ancestral, they are not idiosyncratic.

Both nontraditional and traditional visual arts can have high costs. With the possible exception of historical periods that emphasize genius and downplay technical ability, one has to acquire, through hard work, the skills to create a work of art. The materials have costs. Nontraditional artists purchase paints; traditional artists spend hours finding and then grinding pigments. All artists, it seems, must sacrifice for their art.

Visual art, traditional or nontraditional, has always been used to promote certain behaviors. The traditional arts are unabashedly propagandistic. Traditional people use visual art to encourage traditions that promote cooperation among kinsmen and kinswomen. While it may appear that nontraditional visual art does not attract attention to a message, implicit in this art is, if nothing else, a criticism of the past. Nontraditional artists use art to challenge traditions and encourage new behaviors.

One curious characteristic of traditional visual art is the often cavalier treatment of the art, the final product of the technical process. Elaborate masks can be burned at the end of a ceremony; lovely sandpaintings destroyed. The fact that these objects are said to be sacred, in that the supernatural is said to be involved in their creation and use, does not protect them from this fate. Objects actually made and used by ancestors, such as the *churinga,* were protected as sacred objects. The destruction of these objects was as serious as destroying the tombs and bones of the ancestors themselves.

Nontraditional artists, blue-chip artists at least, sell (or wish to sell) their works for vast amounts of money. The traditional arts were made for local use. In the past hundred years, artists have begun to produce airport art. Based on my eight years' experience working with airport artists in Ecuador, their art differs significantly from art used in rituals or even in daily life. These artists told me that the art made to sell to tourists looked like traditional folk art, but it contained no spirit of ancestors. Important ancestral techniques and motifs intentionally were omitted.

As nontraditional artists became more dependent upon a system of patronage or a market system, they allowed themselves to be influenced by the demands of collectors. Soon, mere craftsmen were said to be distinguished from real artists by their readiness to bow to an influence that the real artists avoided. True art was said to be beyond material or economic considerations. The talk now was that artists created only for themselves. This often was referred to as self-expression. Traditional artists, on the other hand, copied the ancestors and produced art that was in the service of their family and community.

In ancient Mexico, as in other parts of the world, artisans produced visual art objects for the marketplace. The art was made to sell. The higher the quality of the art object, the greater the price that could be asked. While the particular style here could be an ancestral identifier, it was not necessarily so. Such skills could be transmitted from one generation to the next; however, they also could be acquired in an apprenticeship. The art, thus, would be nontraditional. In apprenticeships, artists can go on to compete with their mentors, just as artists today can go on to compete with their art instructors for public attention, sales, collectors, museums, etc.

While traditional artists seem to strive not to attract undue or individualistic attention to their art, such attention is important to nontraditional artists who depend upon their art for income. If an artist can garner significant public notice and acclaim, it is more likely that he or she will be referred to as a genius. Nontraditional visual art, consequently, is more self-interested, aimed at explicitly serving the interests of those making, encouraging, and purchasing the art. As it is self-interested, it is more competitive. As it is more competitive, it is more likely to be subject to rapid change.

The rapidly changing styles of art observed in Europe led philosophers to claim that art could not be defined; in fact, to try and define it was to be exclusive, not inclusive, and to limit its infinite possibilities. While we may limit art by defining it, science and the formulation and testing of hypotheses rest on definitions.

# The Definition of Visual Art
## *The First Step of the Scientific Method*

The whole science of aesthetics fails to do what we might expect from it, being a mental activity calling itself a science; namely it does not define the qualities and laws of art.

—Leo Tolstoy

Evolutionary biologists, who know how important definitions are to the scientific method, do not always feel obligated to precisely or objectively define terms that refer to human cultural behaviors. In a recent manuscript, Randy Thornhill wrote, "some colleagues have suggested that I provide a succinct definition of aesthetics. It is not possible, however, to provide an objective definition based on Darwinian theory" (n.d., 2). Richard Alexander wrote, "I am deliberately vague or imprecise in my usage of a term like 'culture' and 'the arts' because I wish to err on the side of inclusiveness" (2001, 5).

Anthropologists, who fail to agree on many things, including a definition of visual art, are fond of making the claim that *art* is a word rarely used outside of Westernized societies. "It is almost a cliché (perhaps a little too unexamined)," Murphy wrote, "to remark that there is no word for art in the language of this or that people" (1994, 650–51). An implication here is that art is a creation of, and limited to, Western societies. Art's conceptual attachments to fields such as art history or aesthetics makes it an inappropriate term to use for objects found in the prehistoric or ethnographic record (see Conkey 1993; Tomásková 1997; White 1992). We cannot assume, these scholars argue, that the primary function of either prehistoric or ethnographic art was aesthetic. Not only do many cultures lack an equivalent term for art, but they also do not make the aesthetic distinctions that we do.

While the point is well taken that the framework of current art theory is inadequate for explaining ethnographic and prehistoric visual art, does this necessarily mean that the term *art* is not an appropriate word to use when referring to ethnographic or prehistoric objects that we once referred to as art? The term *art,* after all, has ancient roots; some claim that it comes from the Sanskrit word for "making" (Cabanne 1971, 16). The Latin word *ars* means a craft or specialized form of skill (Munro 1949). Throughout the Middle Ages, artists were classified as craftsmen who controlled particular techniques. It was in the sixteenth century that artists were truly distinguished from craftsmen and credited with possessing a particular, individualistic genius. The term *aesthetic* was not coined until the eighteenth century and it is probably safe to assume that prior to that time, an aesthetic emotion was not considered to be a crucial element of art.

Visual art, in other words, has not always had conceptual attachments to what we now call aesthetics. It has not always been an elite activity that is said to be creative and individualistic and capable of arousing a particular aesthetic emotion. While changing assumptions about art's meaning and function are interesting, they do not mean that the term is invalid for the study of prehistoric art, ethnographic art, or even European art prior to the eighteenth century.

Unless we know what visual art is, how can we study it scientifically? How can we identify its origin? How can we determine whether or not it is a behavior found solely in Western societies? How can we know if it has a function? How can we even begin to determine what that function might be? How can we even begin to answer the question posed by Plato and Socrates more than two thousand years ago: What role might art play "in the well-ordered state"? As Tolstoy pointed out in the quotation at the beginning of the chapter, our failure to define art has meant we are unable to agree on very many things.

While definitions are important, it is true that our attempts to understand visual art are complicated by the fact that the term *art* often is extended metaphorically. We hear of the arts of cooking, war, public speaking, and medicine. While these metaphorical extensions may have some logic (all involve mastery of technique), the fact that the term is extended metaphorically does not mean that it has no literal meaning. Are the arts really so mysterious and hard to define? Although philosophers, poets, and some anthropologists can exercise the luxury of claiming that the arts "are too intangible and changing to be defined or classified" (Munro 1949, 5), science should be made of sterner stuff. Before we can attempt to make sense of visual art, we truly do need a definition.

Evaluative definitions, which are more popular in art journals, assume that something is not art unless an art expert has identified it as good art. Good art generally is said to show clear evidence of creativity and intellect and to evoke an emotional response in the viewers. Classificatory definitions, on the other hand, merely ask whether something is or is not a work of art. To classify an object as art does not mean that it has to be a good work of art. According to Dickie, classificatory definitions attempt to "specify the necessary and sufficient conditions needed for something to be a work of art. A necessary condition for being an X is a characteristic which any object must have in order to be an X. A sufficient condition of an X is a characteristic, which, if an object has that characteristic, it is an X" (1971, 41). Socrates argued that if we examine a word's various usages, we will find some element that is common to all of them but not to other things, and we will be able to isolate that element as the essence that defines the category of things (McEvilley 1992, 166). To make a classificatory definition into a scientific one, we add the requirement that the necessary and/or sufficient qualities must be empirically observable.

## Visual Art Is Human Made

Virtually all discussions of visual art make the claim that visual art is man-made; it has a "human creator" (Dissanayake 1992, 29). While I accept that only humans produce visual art, the question this raises is why humans regularly find natural objects (for example, animal art, sunsets, driftwood, colored stones) to be attractive, in that they attract and hold our attention. Further, why do we seem to regularly extend the term *art* metaphorically in order to include these other things? What it is about animal art, sunsets, and colored stones that makes it seem appropriate to refer to them as art? If we assume that there is some logic to this metaphorical extension, perhaps understanding what is common to the metaphors can help identify the term's literal meaning.

Ethologists have been known to use the metaphor animal art to refer to traits, both permanent and seasonal, that can be observed in a number of species. This art includes the brightly colored feathers of birds, the red belly of the stickleback fish, and the red pouch of the frigate bird (Darwin 1871; Diamond 1991). When elephants in the wild use their trunks to make marks in the dust or a stick to make scratch marks on the ground, this behavior can be referred to as art (Diamond 1991). When apes in the wild and in captivity were observed draping themselves with vines and pieces of cloth, Kohler (1925) referred to this behavior as art. The elaborate nests of bowerbirds, woven in a complex design out of hundreds of sticks and, at times, painted with pigments from crushed leaves or oils, are referred to as art (Diamond 1991; Joyce 1975). Satin bowerbirds make a paintbrush by nibbling a piece of bark into an appropriate shape. They hold this tool in their beak to control the flow of a paint solution to decorate a bower (van Lawick-Goodall 1970). Gorillas, orangutan, chimpanzees, and monkeys living in captivity can master painting with a brush or fingers and can work with chalk, crayons, or pencils (Morris 1962). Even experts find this art to be indistinguishable from products accepted as constituting human visual art.

While the frigate bird's pouch seems to be the unlearned product of the developmental process, making and decorating a bowerbird nest involves significant learning. By watching other conspecific males, bowerbirds learn to build bowers that can be as much as nine feet high and weigh several hundred times the weight of the bird. Decorating the bower can involve dragging decorations (brightly color objects, for example) dozens of yards (Diamond 1991).

The implicit definition of animal art seems to specify neither innateness nor learning, nor the expression of a particular emotion. The necessary element of animal art often seems to be the modification of a body or object through the use of form, line, pattern, or color. This decoration can attract attention to a message, including "look at me, or look at this!" Elephants drawing in the sand and primates draping themselves with vines and cloth do not seem to be performing these activities for an audience. As the behaviors do not seem to be noticed, they presumably have no social effect; that is, they do not attract attention nor do they influence the behavior of an observer. They are neither patterned nor predictable.

Emotion is rarely studied by ethologists, as they typically focus on behaviors such as a message sent (an inflated red pouch on a frigate bird) and a response (females notice and mate with the male). Only a few ethologists concern themselves with any emotions experienced by the male or female. Yet, in humans, emotions are a primary element of our studies of visual art. An aesthetic emotion often is said to be a defining feature of art.

## The Aesthetic Emotion

Scholars who define art typically define it by reference to the emotion it is said to arouse in both the viewer and the artists. This definition, while appealing, has a number of problems. First, it fails to distinguish what art *is* from what art *does* (for example, arouse an emotion). Further, it runs the risk of being tautological, inferring a mental state from the art and then using the mental state to explain art (Lewis-Williams 1982). In addition, although the emotion aroused by art is said to be pleasure, much of art is said to arouse grave feelings, or it may leave the viewer bewildered, confused, nonplussed, unsure of any emotional reaction (Anderson 1979). Indeed, art may not arouse any emotion; it may arouse "no aesthetic interest" (Brothwell 1976). Another issue is that "one serious problem with a definition of art that stresses aesthetic or expressive qualities is that such a definition eliminates much of what has been called art for the last seventy years" (McEvilley 1992, 161).

I once bought mangos in the town of Guadalupe, Arizona. As I walked slowly down the road savoring the sweet juice, the only sound I heard was the soft crunch of my shoes on the gravel road. Stopping in the plaza, I studied the simple Yaqui church tucked next to the Catholic one. Moonlight highlighted their whiteness against the dark sky. After a few moments, as I turned to go, blinded temporarily by the darkness, I heard a rustling sound and then found myself, heart pounding, surrounded by dozens of masked and costumed men, dancing, chanting, and beating drums. I was at once both petrified and awestruck. This was the beginning of the Yaqui's Pascua rituals and one of my first experiences with the emotions that can be aroused by art. James Joyce, who was much more of a poet than I, would have described the emotions I experienced as being "apprehended luminously by the mind which has been arrested by its wholeness and fascinated by its harmony" (1928, 480–81). What this apparently means is that the art, the costumes, masks, music, and dance, all exploded into my consciousness; their vividness was in stark contrast to the quiet evening and the small plaza with its simple churches.

I recognize that the emotions associated with visual art and ritual can be profound. I also recognize, however, that not all viewers would have shared my response and that emotions are fleeting and difficult to articulate (Anderson 1979). Further, I don't know what emotion the dancers were experiencing, and I cannot tell what thoughts or beliefs might have inspired their behavior. Are we really safe in assuming that all dancers share the same emotions, thoughts, and/or beliefs?

A more serious issue here, however, is that emotions and mental processes probably exist because of their influence on behavior, particularly social behavior. An exclusive focus on art and emotion may lead us to ignore art's social effects. It is this social effect that is crucial to art's persistence. The assumption that any emotions associated with a behavior imply that the behavior is necessarily adaptive can lead even scientific studies astray. Eating high-fat foods can be pleasurable; eating many such meals could help promote an early death from chronic disease. While the scarcity of fat in our ancestors' diet may have promoted our ancestors' taste for fat, fats are no longer a dietary scarcity. Environments change, and behaviors that were once adaptive may no longer be adaptive. Similarly, although any emotions associated with art may suggest that it once was an adaptation, we cannot use these emotions to argue that it is currently adaptive.

The point of this discussion is not that thoughts or emotions, or even the presumed aesthetic emotion, are irrelevant to visual art. In fact, we can assume that visual art attracts us because it interests us, presumably by provoking some emotion. However, even if we assume that art does arouse an emotion, we still do not know what elicits the emotion. Is it aroused by the color, pattern, form, technique, or the experiences associated with the art object?

## Symbol and Meaning

A widespread scholarly notion specifies that symbols are a necessary if not a sufficient condition of art; an object is art, if and only if that object is symbolic. Influenced by this thinking, Mithen (1998) argues that art originated in large-brained humans who had developed the capacity to make and understand symbols. While symboling may be a marker of our early humanity, and symbols a necessary characteristic of visual art, animals can understand symbols (for example, the sound of the can opener means to the dog it is time to eat). Further, many objects, even natural objects, can become symbolic and have meaning. Wynn, in his discussion of prehistoric tools, writes that tools, "like other features of human culture, do participate in the semiotic domain, and in some very important ways" (1994, 155).

Symbol or meaning, while of interest, may not be a necessary element of art. "Not all societies," as Boas (1955, 88) explained, "have art that is meaningful or has associative connotations." Others apparently agree; Dissanayake writes that the "existence of non-symbolic designs and patterns in human societies suggests that art making is not in any causal or inevitable way dependent on image making or symbolizing" (1992, 90). Even "western art," Charlotte Otten explained, "only partially and sporadically carries this freight of prescribed symbolic meaning" (1971, x).

A focus on symbols, Boas argued, could obscure our study of art. Within a society, he mused, "there can be . . . considerable wavering about the meaning of a symbol"; "in the designs of the Californian Indians, the same form will be called by different people or even by the same people at different times, now a lizard's foot, then a mountain covered with trees, then again an owl's claw" (1955, 13, 102).

The word *symbol,* in common usage, refers to an object associated with and serving to identify something. A problem that discussions of symbol and meaning often fail to address is how a symbol is actually transmitted between individuals; how individuals come to agree that it represents a particular thing. While researchers recognize that "common knowledge is a prerequisite to the functioning of the symbols" (Binford 1971, 16), they seem to assume that the meaning of a symbol is in some as yet unspecified way transmitted from one human brain to another (Coe 1992). Or, perhaps viewers at a "preconscious or even unconscious level" recognize, through participation in a collective unconscious, a particular shape as a symbol of their "deepest aesthetic feeling" (Vinnicomb 1976, 350).

However, unless humans read each other's minds or have a template in their brains that influences a response to a particular symbol, a symbol's meaning implies an identified and remembered association with its referent (Coe 1992). The memory of this association is crucial to the meaning of any symbol; indeed, this association appears to constitute the meaning of a symbol. Meaning, in other words, is learned, and to the extent meaning is shared, it is learned from others presumably through their behavior (speech and actions). As art critic and philosopher George Dickie explained, "the interpretation of symbols in paintings and literature is a public matter. Symbols such as halos have a conventional public meaning similar to the way in which words have public meanings" (1971, 121).

## Creativity and Individualism

Creativity, most art scholars have argued, is a necessary element of art (Alexander 1990; Joyce 1976); some even argue it is a "human need" or "biological predisposition" (Dissanayake 1992, 82). The dictionary defines creativity as "resulting from originality of thought, expression." Creativity, objectively, refers to "deviating from the expected order" (82), or innovation, identifiable change. While many people today see creativity as a necessary characteristic of visual art, Ernst Gombrich, an art historian, argued that "our modern notion that an artist must be 'original,' was by no means shared by most people in the past. An Egyptian, a Chinese, or a Byzantine master would have been greatly puzzled by such a demand. Nor would a medieval artist of Western Europe have understood why he should invent new ways of planning a church, designing a chalice, or of representing the sacred story when the old ones served the purpose so well" (1989, 119).

"The lust for otherness, for newness," wrote Bernard Berenson *may* seem to be the most natural and matter-of-course thing in the world; however, this lust for newness is neither ancient nor universal. In fact,

> prehistoric races are credited with having had so little of it that a change in artifacts is assumed to be a change in populations, one following another. The same holds for the peoples of relatively recent or quite recent date like the Peruvians and the Mayas and Aztecs as well as the African and Oceanic tribes. Even people so civilized as the Egyptians changed so little in three thousand years that it takes training to distinguish a Saitic sculpture from one of the early dynasties. In Mesopotamia also change was slow. But for Alexander's conquest, there might have been almost no newness in India, and but for the Buddhist missionaries as little in China. Why was there so little craving for novelty everywhere on earth? (Berenson 1948, 155)

The words *conservative, traditional,* and *ethnographic,* when used to refer to visual art, do not imply that the visual art will be simple or plain. The visual arts of China and Egypt, along with those produced in many parts of the world, were traditional; most would agree that they are attractive (some might prefer the words *aesthetically pleasing*). This visual art, however, was not creative in the sense of constantly changing or being highly innovative. Magnificent works of art have been produced without creativity and the artistic freedom that creativity implies. According to Hauser (1959, 29), "some of the most magnificent works of art originated . . . in the Ancient Orient under the most dire pressure imaginable [and this proves] that there is no direct relationship between personal freedom of the artists and the aesthetic quality of his works." Artists in traditional societies, in one important sense, are not merely individuals, they are links in a chain going back into the distant past and leading on into the future.

What creativity seems to imply is individualism or self-interested behavior. The practice of using visual art to attract attention to oneself is rare. Among traditional people, one cannot find "an individual style or personal ideals or ambitions—at any rate, there is no sign whatsoever that the artist cherished any feelings of this sort. Soliloquies such as the poems of Archilochus or Sappho . . . the claim to be distinguished from all other artists which is advanced by Aristonothos, attempts to say something already said in a different, though not necessarily better fashion—all this is quite new and heralds a development which now proceeds without a setback (apart from the early Middle Ages) to the present day" (Hauser 1959, 74).

Anthropologists recognized long ago that technical mastery was cross-culturally recognized and appreciated. "Even in the rudest societies," Lowie insisted, "some individuals greatly excel in manual skill so that the most difficult tasks are entrusted to them" (1940, 107). Technical mastery, while not ignored, was not in any obvious way capitalized upon by the technical master. The great painters in China "do not seek honors and wealth, they avoid the climate of the court, and they give away their pictures" (Kris and Kurz 1979, 114). Skill, rather than freeing a talented person, seems to present him or her with more obligations. Those who have mastered techniques not only have to make their own visual art objects, they also have to help others make theirs. Further, technically well-made objects, as Weissner (1984, 204) has pointed out, communicate not only skill but also the diligence, caring, and hard work of the artist. As techniques are a gift from ancestors to descendants, as well as a gift given to ancestors by descendants, diligence is a sign of respect for ancestors. To produce carelessly made art is to disrespect the ancestors.

Our interest in and appreciation of the exotic behavior of many contemporary artists, who are artistic iconoclasts and seem to use their lifestyle to sell their work, does not mean that artists who are not iconoclasts are not interesting, or without passion, or do not produce art that is not attractive. Tonkinson (1978), when discussing the individual behavior of the Mardujara, writes that while most adults were interesting and agreeable people of pleasant disposition, both sexes had a capacity for rapid and passionate arousal of emotions to a violent pitch. Emotions seen most often were anger or sorrow.

"Antisocial and excessive behaviors," wrote Tonkinson, are rare. They occur "for a variety of possible underlying reasons: imperfect socialization, excessive egotism, madness, loss of control, quirks of personality, poorly controlled temper, and so on" (121). Emotions that are likely to lead people into breaking ancestral law (for example, strong dislike, egotism, covetousness, promiscuity and unbridled sexuality, or malicious gossip) must be kept under control. Individuals who break ancestral law are held fully responsible for such acts. It is ironic, perhaps, that characteristics seen as antisocial by the Aborigines (for example, excessive egotism, poorly controlled temper, madness, quirks of personality), have been seen, on and off at least since the Renaissance, as evidence of artistic genius. The primary point here, however, is that madness and creativity, which may be associated with creativity in the expression of an artist's personality, are not necessary conditions of visual art. It is not merely coincidental that egotistical and exotic behaviors demonstrated by artists begin to occur when artists begin to sell their work to demanding patrons.

## An Empirical Working Definition

The characteristics cited by art scholars as being either necessary or sufficient elements of art (that is, the aesthetic emotion, symbols, creativity) are problematic for various reasons. The elements common to all examples of animal art may be a good guide to a working definition of visual art. The modification of a body or object by using color, line, and pattern is also consistent with the characteristics implicit or explicit in definitions of art used by some of the influential thinkers in aesthetics. Plato, who gave us one of the earliest implicit definitions of art, implied that color and form were crucial to art: "I think that you must know, for you have often seen what a poor appearance the tales of poets make when stripped of the colors which music puts upon them . . . they are like faces which were never really beautiful, but only blooming; and now the bloom of youth has passed from them" (1977, 14). In the seventeenth century, Nicolas Poussin (ca. 1647) defined visual art as "an imitation of anything that is to be seen under the sun, done with lines and colors upon a surface" (Goldwater and Treves 1945, 151). Tolstoy's nineteenth-century definition is similar: "To evoke in oneself a feeling one has experienced and having evoked it in oneself, then by means of movements, lines and colors, sounds or forms, expressed in words, so to transmit that" (1977, 65–66).

Early in the twentieth century, Clive Bell, a writer associated with the famous art critic Roger Fry defined visual art as "significant form" (1958, 389).

He claimed that significant form, which included "combinations of lines and color," was the "one quality common to all works of visual art" (18–19; see also Beardsley 1958; Langer 1957). Twelve years after Bell (1958), Boas defined visual art similarly, as significant form. Almost seventy years after Boas, Dissanayake argued that art ("making special") involves, among a vast number of other things, bright colors; appealing shapes . . . and visual contours" (1992, 59). Thus, the following definition is proposed:

> *visual art:* The modification of an object or body through color, line, pattern, and form that is done solely to attract attention to that object or body. Visual art is a mechanism to attract attention to things. As it is used in association with something, it thereby attracts attention to that something. That something may be a message to which visual art draws attention. The proximate aim of visual art is to attract attention, perhaps by provoking emotions. To the extent that visual art is an adaptation, its ultimate function is to influence social behavior in ways that promote success in leaving descendants.

*Attract* comes from the Latin word *attractus,* meaning to draw, or to cause approach or adherence. While *attract* and *attractive* are often used in the sense of attracting only favorable attention, the terms also refer to the fact that visual art draws our attention. The sine qua non of visual art is that it is noticeable. The first requirement of influence is to be noticed. In this chapter, as throughout this book, *attract* and *attractive* are used only in the sense of attracting attention, not of attracting positive attention (Coe 1992).

*Color,* scientifically, refers to the characteristic ways objects have of reflecting various wavelengths of ambient light. For species that discriminate color, what seems significant is that color makes it possible to identify subtle differences in objects. A function of color is to aid in categorization and comparison and in the identification and reidentification of objects (Hilbert 1987). This suggests that colors are attractive to humans because they aided our ancestors in identifying and reidentifying objects and they helped influence appropriate choices (Coe 1992).

*Form* refers to the shape and structure of something, often considered in three dimensions. Humans seem to find some forms more attractive than others. Form aids in the identification of objects, as, for example, fecund females or healthy and strong males (Low 1979; Thornhill 1998). Form also may have been important in such things as landscape classification, evaluation, and orientation for mapping.

*Pattern,* Bateson wrote, refers to "any aggregate of events or objects [that] can be divided in any way by a slash mark, such that an observer perceiving only what is on one side of the slash mark can guess with better than random success, what is on the other side of the slash mark" (1972, 131). Aggregates are patterns when an aggregate's extension can be predicted with greater than chance success. Pattern's importance may lie in the advantages it confers in identifying and reidentifying objects, and thus in making choices (Hilbert 1987; Coe 1992).

In sum, visual art is a behavior that involves making art; the behavior of viewing art is implied. An art object implies the behavior of making art has occurred. One necessary condition of visual art seems to be that humans make it. A second necessary condition is that it involves the use of color, line, pattern, and/or form to modify an object or body. A third necessary condition is that the color, line, pattern, and/or form have no function other than to attract attention, perhaps by provoking emotions. They do not, for example, add structural support to a pottery vessel, prevent dental caries (in the case of tooth straining), or act as a preservative, such as in tanning pelts. Humans have evolved the ability to respond to color, line, pattern, and/or form. Artists, for thousands of years, and often at the encouragement of others, exploit this tendency to respond to color, form, and pattern in order to influence social behavior.

A limitation of this definition is that it refers only to that art involving color, patterns, and/or form: the visual arts. It does not refer to poetry, storytelling, dance, or music, although they clearly involve patterns of sound or of movement that attract attention. Another possible limitation of this definition is that it assumes, but does not focus on, the large human brain or any cognitive processes, or on any emotions associated with making and viewing visual art.

The primary value of this definition is that it is not only in agreement with the usage of the term *visual art* by a number of artists, philosophers, and social scientists but it also focuses on behavior, an objective phenomenon that is potentially measurable. By avoiding creativity, this definition facilitates the cross-cultural study of art, making a dichotomy between contemporary and traditional art, or fine art and craft, unnecessary. Further, as artifacts imply behavior, we can use the term *art* when we refer to objects found in prehistoric sites, ethnographic societies, or in New York galleries. Finally, this definition has the advantage of making explicit an inheritable or replicable unit, a central issue in evolutionary biology.

# CHAPTER 5

# Underpinnings of the Ancestress Hypothesis

Eve . . . was not the only woman on earth, not the most attractive or even the one with the most children. But she was the most fruitful, if one measures fruitfulness by success in propagating a certain set of genes. Eve's genes are in every living human being . . . . she is our 10,000th great-grandmother.

—Brian Fagen

I am not the first, nor will I be the last, to argue that mothers played an important role in the evolution of culture. Shortly after Wallace and Darwin organized their thoughts on evolution through natural selection, Bachofen proposed that the "relationship which stands at the origin of all culture, of every virtue, of every nobler aspect of existence" was "childbearing motherhood" (1967, 79). Over a hundred years later, Briffault echoed this claim, arguing that the "essential foundation . . . of social organization is the direct product of prolonged maternal care and does not exist apart from it" (1931, 57).

Bachofen, and those he influenced, felt that mothers guided "the wild, lawless existence" of our ancestors "toward a milder, friendlier culture" by being self-sacrificing; that is, by encountering "violence with peace, enmity with conciliation, hate with love" (1967, 91). While this sounds melodramatic today, he assumed that harmonious social living was guided by the skills needed to be a successful mother: gentleness, kindness, patience, generosity, trust, self-restraint, and an enduring commitment to the vulnerable that outweighed all other interests. This maternal selflessness apparently was learned: "Raising her young, the woman *learns* earlier than the man to extend her loving care beyond the limits of the ego to another creature, and to direct whatever gifts of invention she possesses to the preservation and improvement of this other's existence" (79, emphasis mine).

Bachofen assumed, as did others of his time, that a mate who, with "desperate valour," defended his home and provided for his children, made it possible for a mother to attend to all her children's needs (Tylor 1960, 151). Before males became assets, however, a mother had to curb male selfishness through her authority over her sons; mothers and wives had "to tame man's primordial strength, to guide it into benign channels" (Bachofen 1967, 144, 151). They had to move adult males into a "voluntary recognition of feminine power": "at times the woman has exerted a great influence on men and on the education and culture of nations. The elevation of women over man arouses our amazement most especially by its contradiction to the relation of physical strength. The law of nature confers the scepter of power on the stronger. If it is torn away from him by feebler hands, other aspects of human nature must have been at work, deeper powers must have made their influence felt" (84, 85).

Although Bachofen argued that maternal selflessness is learned, he felt that paternal care required "a higher degree of moral development than mother love" (79). This was because, as van Baal would explain later, "a man's heart, unlike that of a woman, is at best only partly with his family" (1981, 90). A father's love (and presumably his paternal behavior) was difficult; unlike maternal love, it not only involved learning but also reason. While Bachofen did not describe the reasoning process, he did point out that males as kinsmen and fathers had to learn to restrain their selfishness and use their strength, intellect, and resources for the benefit of vulnerable others. If it is true, however, that human males initially learned to father by watching and copying (and being influenced by) their mothers, then maternal care provided the model for paternal care.

Bachofen's claim that the origin of culture owes a great deal to ancestral mothers, needless to say, did not form the foundation of modern social theory. Although now and then someone focuses on females (Knight 1991), they often focus on her ability to get other females to cooperate in a coalition, or they focus on what they assume is her sole source of power over males: her youthful sexual attractiveness. Few see generous and selfless mothers and grandmothers as influential or theoretically interesting, or even as someone who ever existed.

Although Bachofen's words may seem like little more than wishful thinking, I will be risking little if I build on the argument he raised, as scholars still are uncertain what the origin of culture was or if, when, and/or how humans learn to parent. Biology indeed plays a role in maternal and cultural behaviors. Genes are expressed, however, in an environment that is, for many primates, social. Among humans, mothering behaviors are learned, taught, supported, and reinforced, by and large, through traditional kinship and moral systems (Edel and Edel 1959). Traditions, ancestral strategies from the past, are the key to human parenting. As I assume that the strategies developed by our ancestresses would involve an elaboration of primate behavioral strategies, a clue to the behavior of our ancestral mothers lies there, particularly in behaviors related to maternity, kinship, and lineage or descent. I assume that reproduction for the human female involves more than copulation.

## The Primate Model of Maternal Care

To try to reconstruct the behavior of our distant ancestresses, I focus on Flo, our "dynast" cousin from Gombe, and her children (Goodall 1971, 1999; Hrdy 1999, 30). Flo was not particularly attractive: "Flo looked very old. She appeared frail, with but little flesh on her bones, and thinning hair that was brown rather than black. When she yawned we saw that her teeth were worn right down to the gums." While humans may not have found Flo to be attractive, she was "exceptionally popular" with the males when she went into estrus (1971, 80, 85).

Although it is not clear what made Flo so attractive to males that they treated her differentially when she was in estrus (chasing away adolescent suitors), we cannot discount the possibility that her mothering abilities are related not only to her attractiveness but also to her ability to influence males, as sons and as mates. By calling Flo a dynast, as did Hrdy (1999), we imply that she developed strategies that promoted the success of her descendants, of her lineage through time. Dynastic strategies promote the likelihood that the individuals will become ancestors. Compared with other conspecific mothers, Flo intensified her investment in her offspring, moving toward a greater K-strategy. Flo's mothering behaviors involved intensive care; she spent a great deal of time with her offspring. She was very "watchful" and was "quick to seize [her child] if she saw any sign of social excitement or aggression among other members of the group" (Goodall 1971, 107). She treated her children with affection, demonstrating both "tenderness and patience" (Hrdy 1999, 50), soothing and kissing them when they were afraid (Goodall 1971, 242). Flo also helped promote a relationship between her offspring, who were (possibly half) siblings to one another, by encouraging their cooperation and by distracting them when they engaged in sibling rivalries.

Flo, like other chimpanzee mothers, seems to have slowly transmitted her behavioral skills to her offspring (for example, how to fish for termites) largely by example. Flo taught her daughter Fifi to mother by making certain that Fifi watched her as she cared for Flint, the youngest offspring (Goodall 1971, 106). It is this influence, presumably, that led anthropologists to describe Flo as a dynast.

Flo also was able to continue influencing her sons, even into their adulthood when they helped protect and provision their younger (half) siblings. This assistance made it possible for Flo to devote more time to transmitting the behaviors she had acquired, some possibly inherited from her own mother, to her offspring over a longer period of time. In species with social learning, a lengthy period of time is necessary for mothers to transmit to their offspring the skills the mothers (or perhaps the mother's mother) acquired through trial-and-error learning (see Boesch 1993). The closer the relationship and the longer it endures, the greater the amount of knowledge that can be transmitted.

Not all the mothers at Gombe were like Flo. There were, however, mothers who were inadequate, who lacked Flo's parenting skills. Goodall (1971, 137) writes that these "occasional maternal inadequacies . . . may have marked consequences for the youngsters concerned." A strategy that has marked consequences in one generation can have marked consequences on a lineage. The fact that some chimpanzee mothers at Gombe left no surviving offspring because fate, in the form of a polio epidemic, intervened, does not mean that distinct mothering strategies were irrelevant.

While mothers are at the center of primate social relationships, kinship in some nonhuman primates, such as macaques and baboons, is also "recognized matrilineally" (Reynolds 1994, 139). This apparently means that some nonhuman primates identify descent. There apparently is, within these maternal lines, considerable cooperation. Between maternal lines, or groups of codescendants, there is considerable competition. *Coalition* is the term that primatologists often use to refer to females who form a cooperative group. These coalitions, which are said to center around maternal effort, often form around females referred to as high-ranking. These coalitions can be long term, or stable (Low 2000).

Coalitions, however, are by definition contractual relationships; they form to address a particular problem of mutual concern; they dissolve when the problem is solved. Coalitions involve short-term relationships. While the word *coalition* may be quite suitable for groups of primate females, it may not fit as comfortably with what our ancestresses might have done. As early humans lived in small groups of kin and affines, and as descent categories such as tribes are endogamous, all of these individuals, whether classified as consanguineous or affine, would have shared common descent and thus considered themselves to be kin. The high-ranking females, as well as the allomothers, were kin. While ties between kin, just as ties between a mother and child, tend to be permanent, they are not necessarily so; humans have to work at maintaining even close kinship relationships.

According to Reynolds, the size of the maternal line, health of the individual members, and number of sexually receptive females seem to be the most powerful predictors of the strength of a particular maternal line. This suggests that the males who collaborate with members of a maternal line, rather than just the maternal-line members themselves, help determine the maternal line's relative strength. Sexual attractiveness (that is, being in estrus) apparently is a temporary source of power for a female. While a solitary female is influential only when she is in estrus, a group of females means continuous estrus, with all females benefiting from the sexual advertising of one female after another.

Although Reynold's argument makes the human loss of estrus more intriguing (why did females stop advertising if it gave them such influence?), it cannot explain why older females, compared with younger ones, can have a significant influence on others, not only in some primates but also in traditional humans (Pavelka 1998). It also ignores any influence that females have over sons, brothers, and other kinsmen.

## Different Maternal Strategies

A quick look at the characteristics of ethnographic art indicates that humans, like many other species, are copiers; copying apparently was an important strategy for our ancestors. Throughout much of human evolution, humans apparently copied others who had proven their success: their kin and ancestors. When our ancestresses copied their mothers, they were copying women who, at least in the evolutionary past, had proven they were successful. The fact that humans are prone to copying does not mean that all copied behaviors are equally successful through time. Females can copy an attentive mother's strategy just as they can copy that of a neglectful and cruel mother.

Variation, as Darwin noted, is grist for natural selection's mill. If we follow Darwin in assuming "that any variation in the least degree injurious would be rigidly destroyed" (1962, 91), then we have to face the likelihood that some kinship and mothering strategies will be more successful than others over time. "The vast majority of mothers in eighteenth-century France . . . opted not to rear their own infants but to delegate their care to inadequate wet nurses instead" (Hrdy 1999, 309). Some women apparently provided the care themselves, choosing not to turn their child over to a stranger who was not only inadequate but also did not have the infant's best interests in mind. This example supports that there is variation in maternal strategies and that women can copy unsuccessful maternal strategies, but it raises the issue of which eighteenth-century French mothers became the ancestresses of the present-day French.

Neglectful and cruel mothers are perhaps "likely to be mentally ill, often suicidal, or desperate beyond reason" (Hrdy 1999, 290); if public health data are reliable, the children of neglectful or abusive mothers may be more likely than others to abuse or neglect their own children (Johansson 1987, 90). If it is true that abused children are more likely than nonabused children to abuse their own children, why isn't it possible that they copied the abuse? Why isn't it possible that children of attentive mothers are more likely to grow up to be attentive mothers? Why don't we think of different maternal strategies, serving one's own self-interest versus serving the interests of one's offspring, as having long-term effects, as being competing maternal strategies? Why don't we entertain the notion that strategies that are successful in the short term, in that women conceive and give birth, may not be successful in the long term?

## Factors Promoting Female Reproductive Success

[Human females] are inherently precious—precious by virtue of their biological role in reproduction.

—Robert Wright

Humans, developmentally, are characterized by a long period of juvenile dependency, slow physical growth, onset of sexual maturity that is delayed into the teens, and an unusually long life, even by primate standards (Wood 1994, 14). Among humans, as among other mammalian females, reproduction carries some risks to the mother's health and to her safety (Kruuk 1972; Schaller 1972), probably because of the female's reduced mobility prior to and while giving birth. Pregnancy and lactation may increase a mother's susceptibility to infectious disease (Campbell 1960; Faber and Glasgow 1968). Among humans, pregnant females appear to be more likely to suffer more severely from certain diseases than nonpregnant females— for example, poliomyelitis (Pridelle et al. 1952), pandemic influenza (Freeman and Barnes, 1959), and malaria (Gilles et al. 1969).

While we often equate Darwinian success with success in attracting mates, mate attraction is only a necessary condition. Successful reproduction for the modern human female is a long and complex process involving not only mating and conception but also a series of stages and events: menarche, conception, pregnancy, childbirth, lactation, childcare, and grandchild care. Each aspect of female reproduction involves particular strategies that promote success; however, among humans one factor that appears to be fairly consistently correlated with success is social relationships. Social relationships can provide not only resources and protection but also opportunities for copying.

Further, as we are now beginning to appreciate, the nature of mother-child interactions, beginning in utero and continuing after birth, can trigger hormones that are expressed during the child's developmental process. In an article in *Science News* (November 3, 2001), Michael Meaney's work was outlined. According to Meaney, maternal care influences the brains of offspring. He found that in rats, a mother's behavior triggers her daughters' genetic potential for maternal behavior. Thus, mothering behaviors influence the next generation and, presumably, future generations.

## Menarche, Adolescent Subfecundity, and the Onset of Ovarian Function

Although we often talk about conception as if it necessarily follows from copulation, it does not. Among hunter-gatherers it apparently is not easy for young females who initiate sexual activity at young ages to become pregnant. The !Kung girls reach menarche at 16.5 years but typically do not conceive for three years (Howell 1976). While Montagu (1957), who observed that teenage girls could engage in regular sexual activity without becoming pregnant, referred to this phenomenon as adolescent sterility, these females are not literally sterile, they are subfecund. Unlike sterile women, young females can get pregnant; however, they are unlikely to do so. Human females, like other primates, mature gradually, and there is an appreciable gap between the initiation of reproductive potential and the time of actual reproduction (Altmann 1987).

Skeletal development may be a good predictor of age at menarche (Ellison 1982), as the same neuroendocrine changes seem to initiate both ovarian function and the adolescent growth spurt. Maturation of the skeleton and the switching on of the reproductive system are two separate outcomes of a single set of changes in the hypothalomo-pituitary-gonadal axis (Wood 1994, 405).

Initiation of ovarian function falls near the end of the adolescent growth spurt, and it is at this point that pregnancy is more likely to occur. However, even at this point, a female exposed to regular, unprotected intercourse still may not conceive. Conception, according to Wood (1994) is subject to many vicissitudes: Did ovulation occur? Did at least one act of intercourse fall within the fertile period? Did that act result in fertilization? The probability that a female will conceive is influenced negatively not only by youth but also by increasing maternal age and physical maturity, undernutrition (Spielmann 1989), debilitating diseases (for example, dysentery), acute and chronic infection, severe anemia, and high fevers (Stone 1954; Lawson 1967; McFalls and McFalls 1984).

Not only is it more difficult for young females to get pregnant right after initiation of puberty, but also when they do get pregnant, cephalo-pelvic disproportion (which can occur if the pelvis and its outlet are not fully developed) can cause obstructed labor. Young females, for various reasons, are more likely to experience obstetrical complications, as well as spontaneous abortion and stillbirth. Before antibiotics, young females were more likely to experience puerperal infection, fistulas, subsequent infertility, and maternal death (McFalls and McFalls 1984). They also are more likely to experience severe maternal depletion (a decline in maternal nutritional status that can be attributed to reproduction), as growth is an important competitor for resources (Wood 1994, 16).

If the aim of life is, as we assume, to reproduce, a strong selective pressure against young females reproducing seems curious, as this delay would reduce the number of offspring that a female could produce over her lifetime. Low, who recognizes the complexity of successfully raising a human child, argues that this period of adolescent subfecundity may be adaptive (1998, 134). Girls who matured later had not only greater physical reserves to contribute to their children but also more experience to bring to childrearing, acquired by watching their mothers and aunts and through allomothering experiences. Young females, as described by anthropologists, begin early in life to learn to mother by helping their mothers with caretaking activities (Whiting and Edwards 1988). As their skills develop, responsibilities increase.

## Pregnancy

Adolescent subfecundity suggests that females may be designed to delay the initiation of reproduction, thus reducing the number of offspring they can produce so that they can better care for those offspring. Recent thinking in evolutionary biology, however, does not see this as a sacrifice of potential reproduction. Further, while it is assumed that mothers tend to care for their offspring, the mother-child relationship is seen as conflictual. The womb, rather than a place of protection, is a battleground in which the interests of the mother conflict with those of her developing fetus (see the discussion of David Haig's work in Nesse and Williams 1994). In addition, when rearing a child, the mother's interests conflict with those of her (demanding) offspring.

Bipedal posture, which has its benefits, has costs. The chimpanzees of Gombe, who inspired my hypothesis, do not continually walk bipedally; however, our australopithecine ancestors began to do so more than four million years ago, and that bipedal posture brought with it changes in the human female's pelvis. The pelvis narrowed to hold the intestines in place and facilitate the changing musculature required of bipedal life (Lancaster and Whitten 1990; Shannon 1989). Human females experience an unusually high risk of losing pregnancies, a risk that may be as high as 50 percent (Wood 1994, 14). A cost of reproduction for the female is that fetal tissue growth will be supported at the expense of maternal tissue (Wood 1996, 204). If food resources are limited, sequential pregnancies can result in a cumulative decline in maternal conditions over the course of reproductive life.

Social relationships during pregnancy would have increased the food supply (Feyisetan and Bankole 1991), thus lowering the risk of giving birth to a low-birthweight infant, a condition that is significantly correlated with infant mortality (Spielmann 1989). Further, women with "low psychosocial assets" are three times more likely to experience complications of pregnancy (Boyce et al. 1986, 205), and mothers who experience social stress during pregnancy (for example, fighting and quarreling, death in the family, separation from loved ones) may be more likely to have infants with physical illness, minor physical and functional abnormalities, developmental difficulties, and behavioral abnormalities (Harrenkohl 1986; Stott 1973). In Alto do Cruzeiro, Brazil, almost 15 percent of all single women's pregnancies end with a stillbirth (Scheper-Hughes 1991). Women with social relationships are not only more likely to have a live birth and a healthy child but also to have a low incidence of preterm labor (Boyce 1991), a major determinant of neonatal death and a contributor to problems such as cerebral palsy (Hall 1991).

While we might assume that social strategies aimed at influencing others to invest in us can help a mother reproduce, it may be time to consider that behavior can have multigenerational effects. Parenting behaviors (for example, to nurse or not to nurse) can have multigenerational consequences. If we accept that hormones are triggered when an infant female is cared for, or mothered, and that mothering skills, to some significant degree in humans, are learned, we are left with the fact that if one generation does not mother, the next generation will not have effective mothering behaviors. Although chance may lead to competent mothering and the triggering of related hormones, chance does not account for persistence of cultural behaviors.

## Childbirth

The narrow pelvis of the modern human female brought with it certain "obstetrical challenges and mortality related to birth that is rare among undomesticated animal species" (Trevathan 1987, 22–23). While females can give birth unassisted, the involvement of experienced females in labor and delivery, Trevathan (1987) argues, are of great help. Assistance can shorten labor, decrease the number of complications, and make the experience less traumatic (Bryce 1991). Strategies for assisting in labor while avoiding infection can be transmitted from older females who have had personal experience to younger women. Over the generations, given the large human brain, knowledge about childbirth has been accumulated and honed, and both the knowledge and the relationships that made its transmission possible have become important strategies.

## Lactation

Human milk, like that of other primates, is remarkably dilute and is especially low in fat and protein (while high in lactose) when compared with other mammalian species (Patton and Jensen 1976). As species with dilute milk nurse more or less continuously compared with species with richer milk (Findlay 1974), suckling frequency may have been comparatively high during much of human evolution (Wood 1994).

Nursing a newborn, given the rapid brain growth of human infants, is time-consuming and a tremendous nutrient drain (Nelson and Evans 1961; Wood 1974). Even after the first few months, the energy demands remain high as human mothers in traditional societies continue to nurse their babies almost continuously (Konner 1977). Women in Westernized societies nurse their infants about six times per day, in episodes lasting ten to twenty-five minutes; omitting nighttime feeding within a matter of months is common (Lincoln 1983). Konner and Worthman (1980) found that !Kung women—hunter-gatherers who exhibit features of the social organization thought to characterize those of our ancestors (Lee and DeVore 1976)— nurse their infants about four times each hour for 1.9–1.3 minutes. The Yi-wara of Australia nurse their children until they are four or five years old (Gould 1969). In many respects, this pattern of nursing resembles that displayed by many nonhuman primates. The frequency of nursing and the brevity of the episodes suggest that nursing serves functions, perhaps social functions, other than feeding. Given sequential pregnancies and an extended period of breastfeeding, breastfeeding probably restricted a female's activities for a long period of time.

Both nutritional and social stress experienced by the mother during lactation result in nutritionally stressed infants, as mothers are unable to produce adequate quantities of milk. These infants suffer up to four times greater risk of death than do better nourished children from the same population (Spielmann 1989) because their susceptibility to disease is affected. Janowitz et al. (1981) have estimated that in rural Egypt, children who are breastfed for fewer than three months are about 30 percent less likely to survive to two years of age than those breastfed for twelve months or longer. In populations in which females can choose to bottle-feed their infants, failure to breastfeed is associated with higher rates of gastrointestinal illnesses, respiratory illness, otitis media, bacteremia, and meningitis (Cunningham 1995). There are possible associations of bottle-feeding with paralytic poliomyelitis (Pisacane et al. 1992), deaths from liver disease (Udall, Dixon, and Newman 1985), reduced bone mass in young women (Hirota et al. 1992), and perhaps SIDS (McKenna and Bernshaw 1995).

Dettwyler wrote that "in choosing whether, and for how long, to breast-feed, parents are making decisions that will have long-lasting consequences to their children's health" (1995a: 69). These benefits can be nutritional, immunological, cognitive, and emotional. As Dettwyler explains, that convenience has costs the child pays (1995b, 204). While the majority of our distant ancestresses breastfed their infants, and perhaps even breastfed the infants of kin, it can be predicted that there were different strategies related to length and frequency of breastfeeding and the giving of supplemental foods. Those choices would have had consequences both for the mother, in birth spacing, and for the child (Johansson 1987). These choices also could have had multigenerational consequences.

Weaning, according to parent-offspring conflict thinking, is a traumatic event, with the child trying to manipulate the mother to prolong nursing and the mother trying to manipulate the child to stop so that she can get pregnant again. From another point of view, however, it is in the child's interest to be weaned from the breast in order to adopt a diet that supports the new stages of growth and development. As Super (1981) points out, the trauma may be overrated, as it tends to occur at a sensitive time in vocalization. Like Flo, the chimpanzee mother, human mothers have strategies to divert a child's attention to other things.

Social relationships influence nursing success in various ways. First, as mothers spend much of their time nursing, social relationships may be necessary to help ensure an adequate food supply. Further, maternal choices regarding childcare, including nursing, are affected by "traditional kinship beliefs," which are learned from kin (Dettwyler 2000, 312). If the environment happens to include mothers who encourage their daughters to breast-feed frequently and for some length of time, as well as mothers who urge daughters to do what is convenient, natural selection will act on these traits whether they are transmitted genetically or culturally.

## Infant Care

Human infants are markedly altricial. Human neonates are called extero-gestate fetuses, meaning they are born at a stage of development that other primate infants experience while still in the mother's womb. The human infant is significantly more helpless than any of its primate relatives (Alexander 1990). Phalanges are ossified and many of the organs (for example, the kidneys and brain) and systems (for example, the nervous and digestive systems) mature after birth (Trevathan 1987).

During this early developmental period, finely adjusted responsiveness (Trivers 1974) and mother-child body contact (constant touching and holding, carrying, suckling) appear to be crucial not only to the relationship between mother and child but also to the child's future development. Maternal postpartum depression may have a long-term effect on infants; in some infants, cortisol levels, which are stress indicators, remain elevated for months (*Washington Post* 1999). This early contact also may have an influence on future mothering behavior. In rats, adult females who were frequently licked and groomed by their mothers not only behave similarly toward pups in their care; they also show heightened sensitivity to the hormone estrogen in the brain regions devoted to maternal behavior (Michael J. Meaney, cited in *Science News,* November 3, 2001). The way one is mothered influences how one will mother.

While I hesitate to use the term *social support,* as it is so often vaguely defined, the assistance of others during an infant's developmental period could provide children with better nutrition and give mothers time to spend with their infants and to build social relationships. In Alto do Cruzeiro, mothers who lack such support have babies likely to be "born small and wasted (thin) and many children die in the first year of life as they are . . . weak and passive [with no] resistance against the common crises of infancy: diarrhea, respiratory infections, tropical fevers" (Scheper-Hughes 1991, 145). Higher infant mortality and lack of bonding occur because "children are born without the traditional protection of breastfeeding, subsistence gardens, stable marriages, and multiple caretakers" (143).

## Childcare

Although infants of other species often reach full physical growth and sexual maturity in a short period of time, humans, like chimpanzees, require many years. This juvenile period facilitates the development of the brain and the transmission of knowledge, enabling human juveniles to absorb and adapt to the complexities of culture and human sociality (Alexander and Noonan 1979). During this period, human mothers are limited in their ability to conduct other activities. In traditional societies, childcare may require more than 90 percent of a mother's labor time (Hames 1989).

Because the human offspring's brain is primed to absorb knowledge, it makes sense that mothers would use this time to influence their offspring in ways that would facilitate the mother's role (by getting her offspring to cooperate with her and their siblings) and promote the offspring's survival and future reproduction. Without sibling cooperation, the transmission of traditions is difficult if not impossible. If an offspring does not acquire the appropriate social and subsistence skills, he or she may die.

During childhood, which is the period when children are fed and protected while they learn to be social creatures, parents can influence their children to do things that involve pain. Darwin's father encouraged his son to have an overactive conscience, to be self-critical, to crave reassurance, and to have exaggerated respect for authority. Humans may be "designed to absorb painful guidance that conduces to genetic proliferation (or would have in the ancestral environment)." Therefore, "many things that look like parental cruelty may not be an example of Trivers' parent-offspring conflict" (Wright 1994, 317).

Mothers and fathers, given the nutrient needs of growing children, can find it necessary, as did the !Kung, to invest available resources in their child's growth and development rather than in the maintenance of their own bodies (Lee 1979). Among the !Kung, as among other people, parents go hungry if there is insufficient food. Resource flow is generally from parent to child. A study conducted in Manchester, England, by a twentieth-century physician found that it was possible, based on his many years' experience treating working-class patients, to distinguish two levels of parental investment among the population, all members of which had similar incomes and lived in slum or semi-slum environments (Ashby 1915). According to this physician, recent Jewish immigrants from Eastern Europe demonstrated one level of care. Jewish fathers turned their earnings over to their wives, who "spared no effort to maximize the welfare of their children" (Johansson 1987, 60). English, non-Jewish families demonstrated another level of care. Non-Jewish fathers kept much of the income for themselves, and "the mothers were not so collectively committed to maintaining high standards of cleanliness or medical care" (60). Infant mortality in the area, according to the registrar general of Manchester, indicated that for every 1,000 live-born infants, 150 died during the first year. Infant mortality in Jewish families, however, was 110 per 1,000 births. Given a common set of material constraints (low income and unsanitary conditions), something like 40 of 1,000 of those infants born to non-Jewish working-class families died otherwise avoidable deaths" (60). Johansson concluded that parental underinvestment clearly was not a neutral choice.

Menopause may be, like a long childhood, an important part of the ancestral female's reproductive strategy. In other mammalian species, all of the physiological systems, including that of reproduction (Dunbar 1987), decay at the same rate, and even old females retain some fertility (Smith and Polacheck 1981). In humans, however, reproductive function decays decades earlier than other systems; females at age thirty begin to dramatically lose fertility (Mildvan and Strehler 1960), and it declines thereafter to reach zero between ages forty-five and fifty (Ravenholt and Chao 1974). Human females live perhaps a third of their lives after menopause, while other mammalian females might live 10 percent of their lives after their last birth (Low 1998).

Although menopause may be a by-product of senescence (Rogers and Berntsen 1992), it seems too complex and to come too early in a female's life to be regarded as other than a product of natural selection (Alexander

1990). It has been hypothesized (Alexander 1979; Dawkins 1976; Williams 1966) that menopause became adaptive as the investment in hominid offspring expanded as a consequence of their increased altriciality and dependency. In species in which offspring depend for some time on their mother, it may pay older females to shift from producing more offspring to providing continued high-level care of existing offspring (Cheney and Seyfarth 1990). According to Alexander's Grandchild Altruism Gene hypothesis (1990), it may have been adaptive for an older human female, rather than producing additional offspring, to turn her effort to tending the offspring she had already produced (in order to raise them to maturity before her death) and to aiding her daughters' reproductive efforts.

The relationship between a mother and child is hierarchical. Our conceptions of what a hierarchy is and does are built on our assumption that a hierarchy necessarily involves a "power struggle, in which the powerful often prevail and the weak are often exploited" (Wright 1994, 238). There are, however, as Jane Lancaster pointed out, advantages to a hierarchy. It would be a mistake to "think of dominance as involving uncontrolled or unjust aggression, especially when we think in terms of human relations . . . Dominance is one effective way to organize social interactions . . . unorganized social interactions can be chaotic" (1975, 14).

The mother-child relationship is ranked, but not usually described, as exploitative. As the mother-child relationship is so ancient, a deeper understanding of the characteristics of this hierarchy is crucial to our understanding of human behavior.

## The Maternal Hierarchy

Mammals are distinguished by a ranked relationship; offspring are subordinate to, or dependent upon, a mother who guides while offspring follow. The prolonged immaturity of human and primate offspring reflects not only their dependency but also the mother's responsibility (Steadman 1994). The survival of offspring depends fundamentally on this long-term ranked relationship. The first human hierarchy, or ranked relationship, was that between a mother and her child. Humans appear to respond to such hierarchies and form them often. This hierarchy, which is part of kinship systems in many parts of the world, not only "provides the child with a blueprint for the parameters of most anticipated social interaction" but also allows children to learn "the norms of appropriate kinship behaviors . . . without over-coercion from adults" (Tonkinson 1978, 12).

Mothers have seldom been seen as shedding light on human hierarchies, despite the fact that as early as 1651 Thomas Hobbes recognized that the first human-ranked social relationship was that between a mother and her child. He wrote that "in the condition of mere nature, where there are no matrimonial laws . . . the right of dominion over the child depends on [the mother's] will" (1946, 131). While Hobbes then ignored this ranked relationship, he implied that mothers used their influence to promote the survival of their offspring and to encourage their offspring to cooperate with her and with each other.

Although Hobbes used the word *dominion,* a more appropriate term to use when speaking of the mother-child relationship may be *hierarchy* (Steadman 1997). *Hiero,* the root of the word, is a Greek for "sacred" or "keeper of sacred things"; *archos* means to rule or lead. Hierarchs were leaders of religious groups or societies and were obligated not only to supernatural beings (often ancestors) but also to the people whose servant they were said to be. *Hierarchy,* rather than implying exploitation, may imply generosity, obligation, and even subordination (Santos Granero 1991, 229; van Baal 1981). A hierarch is defined by service, not merely by rights. Hierarchs, like ancestral mothers, are obligated to those who are older, one's ancestors, and those who are younger, one's descendants or metaphorical children.

The association of high rank and duty or obligation is not confined to hierarchs living in the classical world. According to Alfredo Barrera-Vásquez, "Maya hieroglyphic script talks about 'lineage authority' using the Yucatex Mayan term *kuch,* which refers to burden, such as a burden that is carried on a tumpline against one's back, a burden of conscience, a responsibility, an obligation, or the authority of an office" (1980, 343–44). A high-ranked individual in a hierarchy is a servant to low-ranked individuals. To paraphrase van Baal (1981, 114), the higher a person's position in the hierarchy of power, the more is expected, the greater are the obligations.

The exploitation of subordinates, often assumed to be a privilege of rank, is true of a pecking order, not a hierarchy. A pecking order is distinguished from a hierarchy in that the individual at the top has dominance or rank, but no obligations to the one(s) at the bottom, just as the one at the bottom has no influence over the one at the top (Steadman 1997). Pecking orders are impersonal and competitive: hierarchies are personal and involve a vertical form of cooperation. Pecking orders do not imply cooperation in any form: accepting one's fate because one has no choice, or knows that death would be the consequence of making a choice, is not a form of cooperation.

## Hierarchies and Age

The intergenerational relationships among female kin [in Japanese monkeys] represent the most fundamental and enduring relationships in the group. The life-time bonds between mothers and daughters are characterized by affiliative behaviors and by mutual support in times of conflict with non-family members. Because kinship bonds persist across the lifecourse, old females do not become socially isolated, nor experience a decrease in social power.

—Margit Pavelka

It is often assumed that female power comes from the sexual attraction they hold for males. Older females, as they are less sexually attractive than younger females and do not have future sexual attractiveness ahead of them, are often considered especially powerless. This would be particularly true in traditional societies if it is assumed that males use traditions to restrain females. Yet hierarchies often are based on birth order.

Chimpanzees apparently have birth-order hierarchies, as older siblings seem to be obligated to their younger siblings (Goodall 1999). In birth-order hierarchies, which are found cross-culturally among humans, the one at the top of the hierarchy, the older one, is obligated to those beneath her, the younger members of the hierarchy. Among the Chachi, kinship terms depend upon relative age; terms make reference to relative birth position, older to younger (Barrett 1925). Duties fall to the older kin, and deference must be shown to the elder kin. Bird wrote that when traditional Japanese children play a game and there is some doubt about a rule, "instead of a quarrelsome suspension of the game, the fiat of a senior child decides the matter" (1984, 194). When the traditional Chinese made offerings to ancestral portraits, the offerings went down the line, handed from younger to older members, with the oldest then delivering the food directly to the ancestral shrine.

Because humans form hierarchies built on age, it is not surprising that in many traditional societies "advanced age brings increased prestige and authority." In such societies, "senior members of the family and community are elevated to positions of leadership" (Bailey and Peoples 1999, 230). Ohnuki-Tierney claimed that the "aged are the most revered members of Ainu society and are political leaders as well as guardians of religion, oral literature, and other important traditions" (1974, 65). Among the traditional Aymara, a child's taskmistress is an older woman, mother, sister, or mother-in-law who has unquestioned authority (Mitchell 1998, 55). Among the Bemba of Africa, young boys and girls were taught to pay "extreme deference" to age (Richards 1956, 48), and during the rites of passage, girls were told that they were expected "to accept the domination of older women." Rituals opened with a representation of the hierarchy of rank among the women present, with the oldest honored first, then the next oldest, and then on down the line, by exact order of age and rank (54).

Among the Hopi, older mothers (grandmothers) often have significant influence (White 1970, 21). The Irish called old age "the most honored state" (Arensberg 1968, 109). Among the Australian Aborigines, a woman becomes a *kapula* (a respected elder) at age forty-five (Hamilton 1970, 19). Among the !Kung, as in other traditional groups, the elderly were revered as living ancestors (the givers of life); the traditional knowledge they held was seen as crucial for survival (Biesele and Howell 1981, 77–79). Older women among the Netsilik Eskimo are storytellers, and among the Navajo, "if a woman is committed to becoming a chanter she is more likely to do so after menopause," because, at this point in life, she has "more free time to devote to learning all aspects of the ceremonial" (Griffin-Pierce 1992, 39). The Navajo also claim that an older female can take on new, more public roles because she no longer has to worry about infecting her children (if she acquires an illness during a healing), nor does she have scheduling difficulties related to taboos against menstruating women attending rituals.

One role older women can take on is social critic. Among the Inuit, "most men defer to the opinion and wishes of a 'grandmother' as they will to one else," even though those women could be outspoken critics of males (Guemple 1995, 22). The position of these older women in the age hierarchy protect them "from recrimination and retaliation." They can speak openly and even insultingly and "insinuate the 'women's vote' into what would otherwise be an all male 'caucus'" (25). Among the Cherokee, women are influential members of the council who give balance to debate and decisions (Awiakta 1993). Older women often perform another important social role, one that we pejoratively refer to as busybody. They teach manners and use gossip and ostracism to encourage individuals to behave in an appropriate social manner.

While the elders are respected in many societies, it is clear from the evidence seen in our own and other societies that this is not always true. In South America, an old Aché male told Kim Hill and Magdalena Hurtado (1996, as described by Hrdy 1999) that he was a societally sanctioned specialist in eliminating old women deemed no longer useful. He told them about coming up behind an old woman, unawares, and striking her on the head with an ax. While this story may be true, it may be apocryphal, told by a man who had, so to speak, an ax to grind. One has to ask how "no longer useful" was identified and who took care of older, useless males. Did the specialist make this decision alone, or was this a public decision in which the old woman's children and kin had a voice? It seems doubtful that the woman's kin would approve of this action. Briffault wrote that "so important are mothers that even very ferocious people . . . would never hurt their mother because the mother is considered the fountain of kinship" (1931, 128). One has to ask what effect this cavalier attitude toward life, kinship, and the elders would have on social life and on the transmission of the traditions that are, or at least were once, important in defining a people such as the Aché? My guess is that the claim made by this so-called Aché specialist was male braggadocio. The South American Fuegians, Darwin claimed, disgusted him because they ate old women. His account proved to be apocryphal.

Disrespect for elders does exist, and elder abuse is on the rise (Johansson 1987). Most scholars argue that treatment of the elderly is related to their control of resources; elderly females are treated poorly when they no longer are sexually attractive, when they do not control vast amounts of land or money, or when the knowledge they have is seen as obsolete. Elders are valued in societies with no writing system because they serve as repositories of cultural knowledge (see Bailey and Peoples 1999). Elderly traditional Hispanic women, whether from wealthy or poor families, are, however, treated respectfully even though they control no resources and there is a writing system.

Poor treatment of the elderly can be independent of wealth, sexual attractiveness, or a writing system; instead, it is correlated with traditions that encourage us to honor elders and to be socially restrained. This is a testable proposition. Creativity and innovation may often be correlated with poor treatment of the elderly, not necessarily because the elders' knowledge is seen as obsolete but because no one is encouraged to treat the elders with respect.

# CHAPTER 6

# Males as Ancestors

A foundation of modern Darwinian thinking and the assumption underlying sexual selection theory is that there are conflicts of interest between females and males due to the differential investment each makes in offspring (Trivers 1972). Females, by definition, are the sex that initially invests more in offspring as they produce comparatively large gametes that are nutrient rich. Males produce large numbers of gametes, each of which contains little more than genetic material. For humans, the investment disparity between males and females can be particularly significant, as the female's necessary investment is high. Because the net benefits of additional matings are evolutionarily significant for males, it makes sense for males to invest more in mating effort, which in many species involves high risk/high stakes strategies, including male-male contests for mates. Contests for mates can involve high risk of injury, may lead to death, and often result in failure to reproduce at all. Human males, perhaps due to intersexual competition, tend to be stronger and a bit larger than females.

If it makes sense for males to invest in mating efforts as opposed to parenting ones, then the problem is how to explain male paternal investment, or increasing male K-behaviors. In other primates, sons can help care for siblings. This social investment involves risks and decreases the number of offspring a male can produce. Logically, it makes more sense for a male to invest in his own offspring, not those of his half sibling. Further, as the costs paid by the human female to reproduce increase, and child survival depends on her paying such costs, a male's success depends on his sexual access to a K-strategy female. Access to her depends on his increased investment. At this point, an increased investment in social behavior, at the expense of reproductive behavior, would make Darwinian sense. The reproductive behavior of males and females has been highly influenced by traditions aimed at focusing potentially conflicting interests on the interests of the offspring, on the survival, well-being, and future reproductive success of descendants. Marriage, regardless of the form it takes (that is, polygamy, monogamy, or polyandry) is one such tradition.

Marriage, anthropology textbooks tend to agree, is a universal and thus presumably ancient behavior that involves, among other things, a socially recognized and enduring social relationship between a male and a female that is structured around reproductive and economic behaviors. Marriage, thus, is socially influenced mating. Over the years, social science attitudes toward marriage have changed, driven at times by the latest theory or political tide. Sir Edward Burnett Tylor, a father of anthropology, saw marriage as a venerable social institution that defined moral duty: "the mother's tenderness, the father's desperate valour in defense of home, their daily care for the little ones" (1960, 249). Today, however, marriage and families are seen in terms of conflicts of interest between father and mother, parents and children, and siblings. Feminists tend to see this conflict in terms of a male hierarchy and abuse of power.

## Biology of Males and Females: Sexual Selection

George Williams initially described male and female reproduction in terms of the sacrifice that each requires. For males, as the necessary sacrifice is close to zero, there is little to lose and much to gain by demonstrating "an aggressive and immediate willingness to mate with as many females as may be available." The female's sacrifice, however, "may mean a commitment to a prolonged burden, in both the mechanical and physiological sense, and its many attendant stresses and dangers." This large sacrifice means that for a female a poor choice could have long term and costly consequences. Not only could she lose a significant part of her total reproductive potential, but she could produce offspring who, down the line, are unlikely to find a mate, conceive, carry a fetus to term, survive childbirth, or produce and successfully rear her own offspring. Thus, it is in the female's genetic interests to "assume the burdens of reproduction only when circumstances seem propitious" (Williams 1966, 184).

While evolutionary biologists do not refer to males as r-strategists, they often describe them as having the potential to impregnate a different female every night while typically making a small investment of time and energy in each offspring. A female strategy, as opposed to that of a male, is described as representing the K end of the K-r continuum, meaning low fertility, large investment of resources, time, and energy in each offspring. For an r-strategist, marriage is an odd choice. Marriage will reduce the number of mates that a male can acquire and the offspring that he, but not necessarily she, can produce. Males, however, often seek wives; females can be "a permanently scarce commodity in an insatiable market" (Hamilton 1970, 17); in such societies, all females, but not all males, are married.

## Marriage Is Ancient

Geoffrey Miller, who argued marriage occurred late in human evolution, focused on monogamy, or "nearly exclusive sexual partnerships" (2000, 186). Monogamy, however, may be recent. Various lines of evidence support that marriage, while often polygynous, is ancient. Marriage, first of all, is a universal trait; unless it evolved many times in many places, it was spread by early human migration. Second, males interested only in sexual partners, evolutionarily speaking, should desire females at the peak of their fertility (early and mid-twenties) and are most likely to have good pregnancy outcomes. If males just want to impregnate females and leave, they should select these more fertile women. Yet men in many cultures place an emphasis on younger, less fertile females. One explanation for this is that by selecting a young female, a male can possess her long term (Hrdy 1999, 186; see also Symons 1979) and influence her entire reproductive life. If he controls her entire reproductive life, he can end up siring all her offspring, more than he would have sired if he had chosen a female who was even slightly older. This implies marriage.

Female orgasm may also support that marriage is ancient. Female orgasm is associated with a marked increase in levels of oxytocin, the hormone influencing contractions during labor and associated with the "let-down reflex" associated with nursing, as well as the warm sensations felt by a mother when she nurses her baby (Hrdy 1999, 139). While orgasm may be, as Sarah Blaffer Hrdy explains, "an ancient 'maternal' rather than sexual response" and "may have appropriated a hormone used in other contexts," it still is apparently widespread (139). Hrdy grants that orgasm may be adaptive but questions whether it is. Female libido and sexual assertiveness are "dangerous predispositions" for women, as "extra-pair sex" can be risky (223).

However, orgasm and libido do not necessarily promote extra-pair sex; they may help promote bonding in a marriage. A survey conducted by a popular woman's magazine suggests that orgasm is more likely to be experienced by females in enduring relationships (Tavis and Sadd 1977). Orgasm may promote bonded relationships, not only because it dispels tension and strengthens bonds between partners (Hrdy 1999, 223), but also because females who do experience orgasm are more likely to stay with that male (Chesser 1956). Orgasm thus may promote marriage.

The archaeological record provides another, largely unexamined line of evidence supporting that marriage may be ancient (L. Steadman, personal communication, December 1999). The transmission of tool-making behavior in primates (based on impressive circumstantial evidence) is from older to younger individuals, especially from mothers to offspring (Goodall 1973; McGrew 1977, 1992). In hunter-gatherer societies, males, by and large, make hunting tools while women make tools for weaving, preparing skins, and, later, grinding grains. The first evidence of stone tool making in humans dates back at least to the early Homo lineage. While archaeologists do not agree on the function of these early tools, the size and characteristics of a tool such as the Acheulian hand ax suggest that it was a male's tool. If true, then the remarkable persistence of this tool over a million years suggests the transmission of techniques of manufacture and use between male kin. To me, this suggests an enduring social relationship between a father and his son, as a father gains more by investing in his offspring than an uncle gains by investing in a nephew. A relationship between a father and son implies the identification of paternity, which depends on a form of marriage.

## Human Mate Choice

Mating, cross-culturally, historically, and in ethnographic societies, has tended to be socially arranged. The word *marriage* refers to this mating arrangement that limits the choices of both the male and the female. As powerful families that monopolize resources are a fairly recent phenomenon, the arrangement of marriages did not necessarily increase one's power, property, or personal fortune.

While females did not have free choice, females could influence the choice of their mate. Among the Amuesha, an Arawaken-speaking people of the Peruvian tropical forest, "a man cannot give away his daughter or sister in marriage without her previous consent." In fact, "men have little control over their daughters and sisters . . . Amuesha woman are largely responsible for their own decisions, and . . . cannot be forced into taking decisions against their will" (Santos Granero 1991, 232, 238).

Even when females had little influence over choosing their mates, males did not necessarily hold the power. Lester Hiatt (1996) argues that informants from the groups he studied in Australia were firm on the point that a girl's mother had the final say in her bestowal. Annette Hamilton (1970, 19) found that "[o]ld women believe that they have a right to bestow their granddaughters." Jane Goodale (1971), in her book *Tiwi Wives,* claimed that women had considerable power through contracts with the future husbands of their yet unborn daughters. The basic contract among the Tiwi of Bathurst and Melville Islands involved mother-in-law bestowal. In this contract, the male promised to provide his mother-in-law for the rest of her life with goods and services in exchange for taking her daughters as his wife. If he failed to live up to the contract, the mother-in-law could void it, a power not held by the girl's father. Men composed songs not to attract mates but to express the hope that their mother-in-law would not go back on her promise. While males expressed anger at the power the wives' mothers had over them, they did not express it in front of their mother-in-laws; they feared displeasing them.

The aim of ancestral mate selection was to select a mate with whom one's daughter or son could produce descendants who, in turn, would produce descendants. This involved selecting a mate with complementary traditions and skills. Seal hunting skills in a male would not complement the skills of a woman raised in the tropical rainforest. The mate also needed to be physically attractive and likely to protect and not abuse her or their offspring.

## The Woe That Lies in Marriage

Muskogee women lacked access to power, faced onerous constraints, were treated as beasts of burden, and were belittled as irresponsible and immoral. The sexual life of married women was completely under male control. Wives were forced to be monogamous but males could take on additional wives. Males who committed adultery faced penalties "much less severe than that for women." Adulterous women were divorced; the "man and his lover were simply beaten senseless by the offended wife and her female relatives." Even widowhood "did not immediately free women." This male-controlled system forced widows into years of mourning, during which time "they were secluded and remained unkempt and under the control of their deceased husband's female relatives" (Statler 1995, 218, 219).

Richard Statler assumed that males hold power and refuse to relinquish any of it to females. As males are physically more powerful, Statler and others see sexuality as one of the female's few sources of power. Domineering males, by restraining female sexual behavior, rob her of her one source of power. However, it was the male's female kin who monitored the widow's behavior, and the woman's female kin beat the adulterous husband senseless. What Statler describes are females concerned with, reacting to, and exerting strong pressure on the social behavior of others, including the behavior of other females. While he assumes that males use their superior strength to force females to behave, females monitor social behavior among the Muskogee and other people, even those without patriarchal males.

Phyllis Kaberry (1939), attracted by the accounts of early ethnographers who often claimed that Australian Aboriginal males dominated their wives, went to Australia to study how women were treated. While she anticipated finding domineering males, she found the opposite: Australian Aboriginal males did not dominate their wives. She soundly criticized male ethnographers for describing Aboriginal women as victims of male oppression. Far from being chattels and slaves, she found that women had rights, privileges, and complex personalities on par with men. The impression that male anthropologists had of female drudgery and servility was superficial. These males, for example, claimed that women were forced to forage so that males would be free for the drama and joy of the hunt. The reality of hunting, she found, was that it was difficult, generally solitary, and often unrewarding.

Foraging, on the other hand, was leisurely, dependable, and leavened by companions.

While the male ethnographers described males as being cruel and dismissive of their wives, Kaberry found that if a male were mean, his wife would simply pack her belongings and move into a relative's camp for a few days. With no vegetable staples, a male soon came to his senses and sought reconciliation. Hiatt echoed this thinking when he wrote that the Walbiri "could not recall any instance of a married man's eloping with another woman; and it was their opinion that only an insane man would jeopardize his own marriage" (1965, 110).

Kaberry also claimed that Aboriginal women were ritually important because they were linked without bias (meaning they were not second-class descendants) to the same founding ancestors through sacred sites on the clan's land. Clan benefits, as well as the responsibilities to the ancestors and their sites, affected both sexes. While males performed ritual duties on a scale and with a flourish that females seldom matched, this was the women's choice, she insisted. While some females might be impressed by these displays, none of them were overawed. Women also had secret ceremonies of their own that were, as she described them, just as sacred as those of men.

In a similar study, Isabella Bird, who visited the Ainu after hearing the men were domineering, wrote that Ainu men did not like the women to speak to strangers. The women, however, "eat of the same food, and at the same time as the men, laugh and talk before them, and receive equal support and respect in old age." Females' lives, despite the descriptions, were not of "unmitigated drudgery" (1984, 270).

Inuit women also were described by the male ethnographers as being "under the relatively total domination of their male relatives," who had "naturally superior strength and ferocity" (Guemple 1995, 17–18). Early descriptions of Inuit women were "sprinkled with accounts (some of them based on myth and folklore) of the beating, mutilation, sexual coercion, abduction and murder of women; of their hardship during times of shortage; and of their burdens as overworked drudges" (17). In an earlier account, however, Jenness had claimed that marriage among the Inuit "involves no subjection on the part of the woman. She has her own sphere of activity and within that she is as supreme as her husband is in his . . . Both within and without the house she behaves as the equal of the men" (1922, 162). While a male could have the final voice in public decision making, he seldom if ever involved himself in domestic matters. Further, men said nothing although older women were their outspoken critics (1995, 22).

What none of the wives in these examples had was freedom of choice in regard to their sexual behavior. However, to the extent that males were expected to marry, had marriages arranged for them, were discouraged from divorcing, were limited in the number of wives they could have, and retained obligations to their children for life, males did not have complete freedom either.

## Males as Fathers

There is still much debate about the importance of fathers in human evolution. Men show some signs of having been selected as good and helpful fathers, but our paternal instincts have not been well researched yet.

—Geoffrey Miller

When Hobbes wrote that the father has over his children "the *jus vital necisque,* the power of life and death, and *a fortiori* of the uncontrolled corporal chastisement," he implied that the father-child relationship (which was the model for the sovereign-subject model) resembled a pecking order (1946, 133). The relationship between a father and his children, according to Hobbes and those he influenced, was contractual, conflictual, manipulative, and coercive; fathers used their children to serve their father's interests.

Although fatherly behaviors vary cross-culturally (for example, they can be emotionally removed and neglectful or involved and generous), fathers are defined, at least to some degree, by their obligations to their children and their generosity to them (Santos Granero 1991, 262). These social obligations are implied when we distinguish genitors (who contribute sperm) from paters, or social fathers, who invest in offspring (Fox 1993).

Santos Granero criticized the patriarchal father position, arguing that it "stems from an analytical concentration on the duties women have to their fathers and brothers at the expense of an examination of duties men have to their daughters and sisters." He went on to write that "rather than argue that females constitute an economic asset for their fathers or brothers, I shall suggest that daughters and sisters frequently become an economic burden on them" (1991, 232). He echoes this a few pages later, writing that "Amuesha men do not control their daughters and sisters, but have, rather, heavy obligations towards them." Further, males who did not perform their fatherly role properly faced serious consequences.

Santos Granero followed Hobbes in arguing that the father-child relationship is the model for the leader-follower relationship. Power, however, is seen as "legitimate only when the holders of power are perceived as acting in response to the needs of their followers—that is, when they are perceived as ensuring their followers' conditions of existence" and as "generous life-givers" (1991, 229). *Life-givers,* of course, is a metaphor, meaning that males have assumed motherlike behavior; their leadership resembles what I refer to as hierarchical rather than pecking order. Males who did not perform their roles (as fathers or leaders) appropriately faced "justified disobedience" from the people. Justified disobedience among the Amuesha found its expression not in an overt questioning of authority but rather in an indifferent attitude toward it (229).

One factor influencing male involvement in offspring is confidence in paternity, an issue that is difficult in humans because of concealed ovulation (Alexander and Noonan 1979; Bernsdorf and Thornhill 1979; Burley 1979; Hrdy 1979a; Symons 1979). The period of ovulation in the majority of mammals and nonhuman primates is communicated through visual means (estrus bleeding, perianal swelling), odors, or behavior changes. Human females are relatively unique in concealing their ovulation from themselves and from males and in being, at least theoretically, continually sexually receptive, using, some argue, breasts and buttocks as permanent advertisements for receptivity (Jolly 1999).

Based on sexual-selection thinking, two main proposals, referred to by Jared Diamond (1997) as the "daddy-at-home" and "many-fathers" hypotheses, attempt to explain hidden ovulation and extended sexual receptivity. Both theories assume that hidden ovulation is a maternal strategy aimed at getting a father, or fathers, to invest in offspring. The Daddy-at-home hypothesis proposed by Alexander and Noonan (1979) argues that concealment of ovulation would have increased the amount of time a male would have to stay with a female to ensure her pregnancy. Concealed ovulation may be a female strategy aimed at forcing desirable males into consort relationships long enough to limit the males' access to other matings and simultaneously raise his confidence of paternity. This would make expensive paternal investment more profitable.

The many-fathers hypothesis, espoused by Hrdy and others, is equally clever. This position argues that if females live among males so jealous that they actually injure or kill a rival's children, then females can mate so widely that rival males all treat children as possibly their own. Continuous sexual receptivity is a way for females to attract many men, confuse them as to paternity, and receive from them gifts (including, perhaps, good genes) and protection for their offspring.

It is certainly possible, as Alison Jolly insists, that human females "play it both ways" (1999, 186). Possible benefits of extra-pair matings for the female include direct benefits (additional courtship feeding) as well as indirect benefits (for example, good genes from higher quality males, production of a mixed brood with a maximum variety of genotypes, opportunity for mate appraisal in future breeding seasons, and insurance against male infertility) (Westneat, Sherman, and Morton 1990). Despite potential benefits of extra-pair matings, however, there are costs females pay across species. In species in which offspring benefit from nest guarding or in which long foraging trips to partition resources for offspring are necessary, there is the possible loss of paternal investment.

Costs of extra-pair matings can also include retaliation by the paired male if he discovers that he has been cuckolded, as well as an increased risk of parasitism and disease transmission (Westneat, Sherman, and Morton 1990). Further, the logistics of managing multiple relationships might be difficult, particularly given male sexual jealousy and the responsibilities of childcare. Handling multiple mates is a courtship strategy; it is time consuming and can be costly. Further, a new male would be a threat to a woman's existing offspring, as he would know that he could not possibly be the father.

Males also pay costs across species. While the extra-pair male faces harassment and possibly injury, the cuckold faces possible retaliation from the female's new mate. While a male benefit may be additional offspring, this is not necessarily true. Schwartz et al. (1999) found in their study of Humboldt penguins that none of the extra-pair copulations resulted in fertilizations; all of the offspring born to females who had extra-pair copulations were attributed to the putative father. Schwartz et al. propose that the extra-pair mating was either epiphenomenal, or else it was aimed at finding a mate for future seasons. None of the females, however, switched mates before the next breeding season. Across species, males who invest in offspring spend less time seeking extra-pair copulations (Birkhead and Møller 1996; Westneat, Sherman, and Morton 1990).

While the costs and benefits could be examined forever, the important fact here is that logically there will not be a selection for fatherly behavior unless fathers are caring for their own offspring. While Miller proposes that "human fathering instincts may have evolved through sexual selection for pleasing the existing children of potential female mates," multiple unions would select for Don Juan's genes, not fatherly genes (2000, 194). Males are likely to limit their investment in a female and her offspring unless they are assured (through her behavior) that they are indeed the fathers (Alexander and Noonan 1979; Blake 1955).

While it must be true that a "mother's goal . . . is generally 'quality,' well-spaced, healthy offspring, each one provided for" (Hrdy 1999, 255), it may be true that paternal care, or at least a permanent mate, makes a difference in whether or not this is likely to occur. In her study of sexual unions in Jamaica, Blake (1955), found that fathers did not want, acknowledge, or support children who were the result of purely sexual interest and that the survival of such children was appreciably lower than for those offspring with a father. Even though a male might indicate that he was interested in having a baby with a current girlfriend, this did not imply he would stay interested in that child when the relationship ended. Twenty-three percent of broken unions were dissolved at some point in pregnancy, caused by the fact of pregnancy.

A pattern of multiple unions (male polygyny/female serial monogamy), when compared with monogamy, was associated with decreased fertility. Blake estimated that the loss of fertility, due to multiple unions, was one-third to one-half. Women involved in multiple unions spent a considerable period of time between unions when they were not exposed to pregnancy. Further, the most fecund women were withdrawn from the breeding population. Women spent the largest proportion of their nonunion time (after sexual maturity) prior to age twenty-five, the period of their highest rather than lowest fertility.

Blake concluded that the principle of legitimacy plays a crucial role in reproduction. This social rule specifies that a child should not be brought into the world without a father, one man who is a guardian and a protector and the link between the child and the rest of the community (1955, 32). Because the father-child bond is so weak biologically, the family cannot exist without that universal rule, as neither laws nor institutional structures are strong enough to give a male compelling interest either legally or emotionally in all his offspring born of a short-term relationship.

Marriage makes the identification of fathers possible. Not only are offspring linked with fathers, but also, through that father, the offspring are linked with a large number of male kin. Marriage, however, means that human males are moved from a strategy of attracting many mates and siring many offspring, an r-strategy, to a strategy more like the female's, producing fewer offspring and providing more care to each.

Marriage, as well as the tie between a father and child, apparently appears to profit from some cultural manipulation. Among some Amazonian tribes, fathers are involved in a tribal ritual celebrated at the birth of a child. By painting the small child with red annatto, a father acknowledges paternity and accepts the obligations that attend that title. Single motherhood, in traditional societies, is regularly discouraged. Among the Bemba of Africa, a child born to such a mother is called "a creature of ill-omen—*wa mputula*—who would bring misfortune on any village in which it lived. Such a child would make it stop raining; granaries would empty quickly; there would be widespread dissention" (Richards 1956). To prevent these negative consequences, the child's mother and father (who would apparently be identified, even if he did not identify himself) would be driven out into the bush. This behavior undoubtedly seems harsh, particularly in a society that values its pleasures. If one asks, however, what might be the effect of such traditions, one possible effect is that it would discourage childbirth outside marriage. It would reduce the chance that a child would be born without the social relationships that may be crucial to its survival (Blake 1955), future reproduction, and its ancestors' dynastic success.

Marriages, while they tend to be enduring, can and do end before the death of either partner. Making a good mate choice is particularly important when marriage is regarded as a permanent relationship, which in most traditional societies it is. The Walbiri of Australia "believe that the marriage tie and the rights of spouses should be protected from injury wherever possible. Permanent, stable unions are the ideal, for these ensure that not only will more children be conceived, but also that the children will be reared properly—that is, as good Walbiri. Separations and divorces are in fact comparatively rare" (Meggitt 1965d, 103). In other words, divorce and separation are discouraged, perhaps because of the effect that they can have on subsequent generations. The breakdown or loss of important social relationships can affect birth outcomes, as well as a child's nutrition and risk of abuse or injury. Researchers have noted that children under ten who living with stepparents are between thirty and forty times more likely to suffer parental abuse (see Daly and Wilson 1988b). Divorce also often means that a child will be raised without a father, and thus will not have the opportunity to learn the skills to be a father, a spouse, or a grandfather. Divorce can affect generations of the family.

Santos Granero noted that divorce and separation also can have an effect on many of the cooperative relationships that are so important to humans. They are "always disruptive for all those concerned, and the breaking of a marital link always implies the interruption of a whole chain of exchanges between two groups that are associated by links of repeated affinity. Unless there are serious reasons for a divorce, a woman's father or brothers, will therefore generally try to persuade both parties to be reasonable and reach a peaceful agreement" (1991, 239).

## Females as Wives

An issue related to male dominance is the restraint of females and female docility and coyness. As *docile* refers to "tractable" or "easy to influence," it is at times assumed that a docile female is a male's plaything. Female docility, or her ability to be influenced, may indeed serve male interests, just as docility and readiness to be influenced serve a mother's interests and, presumably, in the long term, a child's interests. It certainly seems to be true that males in Western societies seldom criticize (except perhaps in the last few decades) females as being too docile. A more typical criticism has been that females are hysterical or too emotional and thus nonrational. What this criticism may really mean is that the female is not easily influenced by the male and that she is, in fact, trying to influence him.

As docility has often been traditionally encouraged, the issue is how it might have served our ancestors' long-term interests. Coyness in post-menopausal females, as novelists have long noted, is ludicrous; coyness belongs to youth and perhaps to an association with courtship. Young Inuit women "are expected to be coquettish and, therefore . . . appear reluctant to marry" (Guemple 1995, 23). Docility, thus, could slow down a male's courtship while promoting male generosity. It would facilitate a female's choice of a mate by giving her (and her parents) time to measure a male's interest and mettle.

Another possibility, however, is that given the greater strength of males, one would expect females to have little control over their choice. However, a female's control over her choice is critically important, as the investment she makes in offspring is large. Males who are unrestrained, sexually and competitively, will always be a threat to a female's choice. Female choice, if this is true, should select for more docile males. Restraint of male sexuality and competitiveness (through such things as marriage) is largely encouraged through traditions. Males, however, will not long be restrained if young females are unrestrained. In fact, one fairly easy way to restrain males may be to restrain young females (who are more social and thus perhaps easier to influence socially) and to use traditions to limit a male's access to young females.

Traditions encourage mothers to put their children's interests first, to avoid males interested in short-term relationship, and to get others who share an interest in the child's well-being to help provide the abundant resources and constant care and attention needed by children. One of the relationships that traditions regularly encourage women to form is an enduring relationship with a male who is father to her offspring.

# CHAPTER 7

# Visual Art Theory

## *Ancestress Strategy or Sexual Strategy?*

From ancient times to the present day, the great bulk of the theories of art have been functional.

—George Dickie

For thousands of years, humans living even in the harshest environments have managed to decorate a multitude of objects and have used significant amounts of resources to do so. It seems curious, given the assumption that humans regularly conduct cost-benefit analyses, that time and resources would be devoted to the production and viewing of visual art, an activity that many see as peripheral and even frivolous.

An aim of visual art, as it was used in the ancestral past and is used by traditional people, was to identify kinsmen and codescendants and to temper aggression and competition in order to promote kinshiplike cooperation among those so identified. The effect of art that promotes its replication through time is the creation of an environment in which large numbers of individuals who identified themselves as kin and codescendants are the kin, teachers, protectors, and providers of resources for costly and vulnerable human offspring.

While Geoffrey Miller's book *The Mating Mind* (2000) has been criticized for a number of reasons, no one has really damaged his assumptions, namely that visual art, at some level, is about competition, either between groups or between individuals. While I accept that competition is involved, the competition apparently originated long ago, when one of our distant ancestresses developed a dynastic strategy that involved (1) limiting the number of offspring that she produced and that each of her descendants produced while significantly increasing the investment made in each offspring and (2) using visual art to identify kin and codescendants and encourage their cooperation. In each generation, individuals chose whether or not to respond to the maternal (and then paternal) hierarchy and copy behaviors they observed. The environment for those who accepted the strategy (the traditions) was one in which a large investment in offspring was less of a risk, as those costly offspring were protected and nourished by others who claimed kinship. The strategy must have worked in the long term, even though in the short term this mother and each of her descendants was outreproduced by less-involved mothers.

Miller, on the other hand, argues that while art is not aimed at promoting maternal competition, it is competitive; it evolved as a courtship tool and functioned, "at least originally, to attract sexual partners by playing upon their senses and displaying one's fitness." Miller suspects that early forms of visual art are about sex and competition for mates. I argue that they are maternal, in the sense of sacrificing one's own self-interest, to promote the interests of another person. Miller (focusing on number of mates) believes art is a courtship strategy that is, through competition, driven to exaggeration. Focusing on the costs of reproduction, I argue that it is a maternal strategy that places limits on one's ability to use visual art to promote one's own self-interests. He asserts that art is aimed at influencing a series of temporary mates; I argue that among traditional people, visual art is aimed at influencing enduring cooperative relationships with children, kin, and permanent mates.

I see older females, who tend to be the scrutinizers of social behavior, as a source of visual art; Miller seems to find no place for them in a world of active libidos. Miller denies that art was used for propaganda purposes. I argue that propaganda, or the use the arts to attempt to influence certain behavior, is precisely its aim in both traditional and nontraditional societies. Visual art regularly has been used to encourage, among a very large number of other behaviors (for example, notice this person; read this book about Dick and Jane): moral, patriotic, religious, and, more recently, purchasing (for example, buy this brand of cigarettes or that brand of single malt scotch).

I see religion, specifically ancestor worship, as playing a dramatic role in evolution of visual art and social behavior; religion and religious functions for art, Miller claims, make no evolutionary sense. Miller grants importance to persistence solely because the persistence of style provides a model or template against which comparisons of excellence and creativity can be made. I agree that traditional visual art is the template; however, I argue that in traditional societies, careful and accurate copying is more important than a creative edge.

Although our two hypotheses seem to provide a theoretical balance, male versus female strategies perhaps, Miller's proposal can explain only a small portion of the visual art produced by humans—namely visual art produced in particular historical periods such as since the Renaissance— or that is found among people, such as the Greeks, who abandoned their ancestors. The ancestress hypothesis, in contrast, not only describes the characteristics of traditional (conservative, kinship-based, largely ancestral, cooperative) and nontraditional visual art (quick to change, competitive, individualistic, creative, at times sexual), but also predicts the conditions under which either is apt to arise. When ancestors are honored and their traditions continue to be both important and transmitted between generations of kin, visual art is conservative. When traditions and ancestors are lost, kinship systems are weakened and the arts set forth on new paths that emphasize individual artists and their particular skills.

## The Ancestress Hypothesis: The Basic Model

The ancestress hypothesis proposes that survival and reproduction, while necessary conditions for humans to leave descendants, are not sufficient conditions for so doing. Social behavior also appears to be a necessary condition, even though it is at the expense of both reproduction and survival. If we are to understand the traits that characterize our species—not only our complex culture but also our extreme intelligence, adolescent subfecundity, exterogestate fetuses, extended period of juvenile dependency, menopause, support of reproduction by older postreproductive individuals and by males, and a long life span—we must understand human patterns of kinship and the maternal hierarchy. These patterns of kinship center on mothers, grandmothers, and children and involve the identification of and cooperation between close kin and codescendants, which are encouraged by traditions.

High levels of knowledge, cleverness, coordination, and cooperation are necessary if a mother is to provide skilled care for the increasingly altricial and slow-developing offspring who must be reared in a world that is socially complex as well as dangerous, particularly for a child with very few or no natural defenses. An increasing need for information storage and processing, particularly strategies for promoting the enduring social relationships crucial to the transmission of complex behaviors, was a primary factor promoting the evolution of the large brain. Longer juvenile periods would have facilitated extended periods of learning, which were needed to acquire appropriate skills, including complex social skills. The arts would have made the message more attractive and more memorable. Intergenerational ties would be a product of the demand for an increased flow of knowledge from old to young. As wisdom and knowledge accumulate as one ages, selection would favor lowered mortality rates and greater longevity.

## Visual Art as Ancestress Strategy

The assumption that underlies the ancestress model is that the linchpin of modern human biology and behavior, including cultural behavior, is the self-sacrificing K-strategy ancestral mother who increased the investment she made in her offspring, even though her more self-interested r-strategy sisters outreproduced her. Further, not only did this ancestress increase the investment she made in her own offspring, she also encouraged her off-spring to increase the investment they made in their offspring, even though that investment was also at the cost of their survival and reproduction. In what is now outdated parlance, such a mother is referred to as a good mother.

The best way to portray this good mother is to describe her hierarchy, which is characterized by the obligations that the one at the top has for those beneath her. Crucial to this dynast's strategy was the use of her influence to promote the involvement in childrearing of alloparents, male and female kin who helped care for the children, and the long-term commitment of her mate and father of her children. Also crucial was the use of her influence to encourage cooperation among her offspring. Without these forms of cooperation, neither her increasing investment nor the transmission of complex traditions would be possible. Enduring social relationships rather than short-term mating ones, generosity rather than selfishness, and cultural strategies to promote confluence of interest rather than conflict of interest would have characterized the social relationships among these early human dynasts.

As the maternal hierarchy is central to the dynastic strategy, age carries the responsibility of preparing the next generation and youth carries the responsibility of acquiring behavioral competencies or technical and social skills. Central to the transmission of skill is respect for the elders, from whom the skills come, and responsiveness to the concerns of youth, as the young vary in such skills as quickness and dexterity.

Visual art as an ancestress or dynastic strategy is highly social. It is used not only to identify kin and promote kinship cooperation but also to promote the replication of strategies related to maternal concerns: conception, a safe pregnancy and childbirth, successful infant care and child care, and kinship behaviors, including care of the sick and rituals for the dead, who, after all, were once kin and are now ancestors.

If it is true that visual art is social, then it is not a solitary activity or a competitive or skeptical one. Skepticism and competition are always at the cost of cooperation. Further, dynastic visual art is aimed at promoting enduring social relationships, which are crucial to the transmission of complex behaviors, not at attracting short-term ones. Short-term relationships can be dangerous, particularly if they are associated with sexual jealousy. Increased rates of homicides, infanticide, child abuse, suicide, domestic violence, and rape are related to sex and sexual jealousy (Thornhill and Palmer 2000).

The origins of visual art could have been accidental. Perhaps it had its origin when a mother adorned her children in a certain way, maybe arranging their hair in a particular fashion. If her children copied that hair arrangement and they encouraged their own offspring to copy it also, then the hair decoration would identify the descendants of the original ancestress. Over time, given differential reproductive success, more and more individuals could have come to be identified as codescendants. If the visual arts were also associated with a message about the ancestress and her behavioral expectations, namely to be generous to and cooperate with those identified as kin, then we would expect to see an effect. What art and its related traditions could have created is an environment in which large numbers of individuals, identified as kin, were the defenders and providers of fragile, costly human offspring.

While one would not expect to find that females used dynastic visual art to attract a series of mates, male decoration, given the importance of female choice, would not be surprising. Male decoration would give a female and her kin information that is important in making a mate choice. His inherited decoration would communicate his ancestry and his commitment to his ancestors and codescendants. The quality and, dare I say it, conformity of his decoration would show his cooperativeness and the size of his kin alliance, and it would tell her whether or not his traditions complement hers. Such information could have been crucial in making a mate choice.

Visual art as a dynastic strategy, as a social strategy, is used to encourage kinshiplike generosity, a trait that characterizes mothers and is, as Santos Granero (1991) argues, a crucial aspect of cooperation. Tonkinson (1978, 120) writes that among the Mardujara Aborigines, it is felt that before a child can walk or talk, he or she should learn about sharing with others. Just as Anna Shaw learned generosity when she gave away her first finely woven basket, the children of the !Kung San of the Kalahari learn generosity when they give away a colored bead necklace they received when they were under six months old. When they are one year old, the grandmother cuts off the beads, puts them in the small child's hands, and encourages that child to give the beads away to an older relative (Weissner 1977).

Visual art as a dynastic strategy would encourage restraint of behavior, which as Jane Goodall (1999) argues, is essential to complex human social life. Restraint, in the case of dynastic visual art, involves conservatism, meaning that limitations are placed on one's self-expression, on one's ability to use visual art to increase one's own individual prestige or power. Tonkinson (1978) argues that restraint is expected of all mature Mardujara males. "A man's behavior toward most consanguineal adult kin is marked by various degrees of restraint" (47).

Because traditions regularly encourage restraint, it apparently is learned and must be encouraged. I would follow Bachofen in arguing that the human female was the one who, through childcare and kinship activities, first learned restraint. Males among some primates learn restraint, as Flo's sons learned from their mothers. Berner (1992, 41) explains that humans learn restraint by copying. Boys, he claims, learn these behaviors through imitation, by watching adult males who explicitly encourage these behaviors. Older Mardujara males teach restraint to initiates by telling them "Keep your minds on the Law, not what hangs between your legs!" (Tonkinson 1978, 101). If it is true that self-restraint involves "a long process of learning" (Newberger 1999, 163), then enduring social relationships would be important. If social relationships tend to endure, rather than being short term, people should tend to be more conservative and more restrained. If one rejects a tradition, one risks damaging the social relationships that were formed during the learning process and in which one has already made a considerable investment.

Perhaps the next question is how does one distinguish between restraint encouraged by one at the top of a pecking order? The hierarch is interested in the well-being of the lower-ranked individuals, restraining their behavior in order to protect their best interests. The interest of a hierarch, just as the interest of a parent, is in keeping his or her child out of harm's way, in promoting their survival and reproduction even if it is at one's cost. This is true in the case of one's daughters, who could pay a high price if they behave foolishly, and one's sons, who are likely to take high risks. The head of a pecking order, on the other hand, is impersonal and interested in his or her own well-being. A female's foolishness and a male's bravado are used to one's advantage. In sum, encouraging restraint in subordinates is at the cost of the hierarch's survival and reproduction, as it is time and resource consuming. Restraint related to a pecking order, on the other hand, benefits the survival and reproduction of the one at the top but at the cost of the survival and reproduction of the lower-ranked individual. Being social, as opposed to being impersonal, thus defined, has survival and reproductive costs.

Although most studies of culture are founded on theories of mind, which propose that the human mind evolved as a survival tool, I agree with Miller that we are far more intelligent, entertaining, social, and articulate than the demands of surviving on the plains of Pleistocene Africa would require. Contrary to Miller's thinking, however, humans, cross-culturally, are also more conservative, hierarchical, and social than the demands of finding a temporary sexual partner on the plains of Pleistocene Africa would require. Traditional art is produced at high costs, thus threatening survival, and has limitations placed on it that inhibit its use for mate attraction. By intelligently and carefully selecting an intelligent and careful mate, and intelligently and carefully encouraging enduring, cooperative behavior among a large number of individuals identified as kin (thus building and maintaining extended kinship networks), our ancestresses, with their K-strategy, were the force behind the evolution of modern humans and their brains.

## The Mating Mind: The Basic Hypothesis

Like any nouveau-riche connoisseur, we are both proud of our art and ashamed of our ancestry.

—Geoffrey Miller

On the face of it, Geoffrey Miller's argument is a simple one. While biologists often describe natural selection as a process that selects the better survivors and reproducers, Darwin himself distinguished natural selection, which promotes survival, from sexual selection, which promotes reproduction. Miller (2000) draws on this distinction to argue that our unrelenting search for survival benefits to inherited traits underlies our inability to understand some of humankind's most intriguing characteristics such as art and music. Survival functions have yet to be identified for many of these traits, Miller explains, because they are tools for attracting sexual partners.

Sexual selection theory, first conceptualized by Darwin, is built on the assumption that there are significant differences, sperm versus egg, between males and females in all species and that these differences influence behavior. Females, who initially invest more in offspring, benefit from making a careful mate choice because they have much to lose if they mate frivolously. In theory, a male (if he could lure enough females), can reproduce once or even several times every day. Males who attracted and mated with as many females as possible were more likely to become fathers. While concealed ovulation complicates things (human males have to stay around a while if they are to get a female pregnant), it is assumed that females will devote their time and energy to growing eggs, being pregnant, and producing milk and caring for offspring while males use that time and energy "for reproductive competition and courtship" (87).

Selecting the best mate required "perception, cognition, memory, and judgment" (8); however, by "intelligently choosing their sexual partners for their mental abilities [our ancestors] became the intelligent force behind the human mind's evolution" (4). The best mind, needless to say, was the more entertaining one: the "human mind's most impressive abilities are like the peacock's tail; they are courtship tools, evolved to attract and entertain sexual partners" (4). This peacock tail of a mind, however, is produced at high costs and not only does it influence such obvious means of attracting mates, such as decorating one's body, but it influences humor, storytelling, gossip, art, music, poetic language, ideologies, religion, kindness, political convictions, and creativity (15, 18, 97), all of which can attract mates. Any excess, apparently, can be aphrodisiac.

Miller borrows from Fisher's runaway selection hypothesis to explain the positive feedback loop, with female choice driving male ornamentation.

Runaway selection, however, cannot account for the patterns observed in brain development, and art, for that matter (bursts of development and then periods of stasis), nor can it account for elaboration of ornamentation. Runaway selection has no bias in any evolutionary direction; it could select for drab and dumb as easily as it could select for conspicuous and brilliant, as long as drab and dumb are considered sexually attractive.

What solves the problem of the neutrality of runaway selection is the idea that runaway traits, whether observed in males or females, can become fitness indicators, communicating that one is an appropriate mate. Miller borrows from Zahavi, who argued that many signals, including sexual ornamentation, evolved as advertisements for an animal's fitness. According to Zahavi (1975), ornaments have high costs because the high costs are what keep the ornaments reliable as indicators of fitness. "Waste," Miller argues, "is what keeps the fitness indicators honest" (128). The antlers of Irish elk and the tail of the peacock require lots of energy to grow and display, energy that unfit, unhealthy animals do not have. This condition-dependent thinking led Miller to conclude that "the only reliable way to advertise one's fitness is to produce a signal that costs a lot of fitness" (123).

The evolution of art, Miller argues, involved the evolution of a human tendency to make material objects into "advertisements of our fitness" (285). Art, like spiderwebs and bowerbird nests, is an extended phenotype, with the genes reaching out from the body into the environment. Artists (variously defined: wordsmith, storyteller, comedian, painter, politician, sculptor) communicate their fitness by making something that has costs that are so high (involving, if nothing else, complex wiring in the brain) that lower fitness competitors cannot imitate it. Aesthetics, or a sense of the beautiful, refers to "human preferences that evolved to favor features of man-made objects that reliably indicate the artisan's fitness" (285).

Miller's hypothesis is clever and has attracted a great deal of attention; however, as H. G. Blocker argued in his *Philosophy of Art,* a theory does not have to be correct to have great historical impact (1979, 170). The validity of Miller's hypothesis rests, of course, upon the strength of the evidence. While, on a fundamental level, it is not clear that sexual selection drove the evolution of the human brain, on a more testable level, I have raised some issues to consider. To some large extent, Miller's hypothesis rests upon serial mating, with both males and females spending much of their lives in mate-attracting activities. Not only is marriage apparently ancient, but this assumption also ignores the tremendous investment that our ancestors made in their offspring—an investment that would have been at the expense of mate seeking. Miller's hypothesis also rests upon cultural excess and exaggeration. How then do we explain long periods of stasis and the periods of history when humans turned away from excesses and back to simplicity? The baroque and rococo periods do not tell the whole story of art. How do we explain visual art's ties to kinship and ancestors? How do we explain why both the strong and the weak are dressed similarly? Unless we fall back on mindlessness (which Miller does) or coercive males (which, thankfully, he is not so prone to do), how do we explain traditions, which limit one's self-expression? How do we explain the self-restraining that is found in hunter-gatherer societies and that presumably existed long before the earliest hierarchies appeared? Each new hypotheses raises new questions, and for this we owe Miller a debt of gratitude.

## Issues Related to Ancestors and Mating Success

### Mothers and the Arts

Miller's primary focus is on what mothers tend not to be: exotic, noticeable, brightly colored like peacocks. Children, not courtship, are "every mother's paramount concern" (193). Mothers, Miller argues, balanced their mothering activities with mate-attracting ones, turning normal motherly duties into fitness indicators that would entice males. By embellishing their child-care activities, not to make those activities better for or more attractive to children (although that may have been a side effect), a female could attract a series of males.

Miller seems to imply here that art for children would be second rate, a mere side-effect of using art for mating purposes. By making the story appeal to any adult males in the vicinity, mothers could attract those males as mates (193). Did mothers make art for their children when males were gone? Did females make art when other females but no males were around? Did postmenopausal females make art? Yes, they did. A more critical issue here, however, is not only whether a mother with a wandering eye (wandering in order to see if there were any attractive males in the vicinity and whether or not, and how, those males were responding to her art) would lose the opportunity to teach her children while telling a story, but whether or not mothers focusing on such things while simultaneously caring for her young children would risk having the children wander off. An investment in courtship can be at the expense of an investment in mothering.

In my fieldwork in Ecuador, it seemed clear that when mothers invest in courtship (purchasing makeup, perfume, and clothing, visiting beauty shops, spending time in bars) it could bring a return. While my data is anecdotal and based on only five single mothers, the bulk of the resources that such behavior brought in were spent on purchasing more beauty products, not on producing offspring. If there was a grandmother, she raised the children. Mothers without such resources left their children alone in the home. When these mothers brought males home or took their children into a male's home, the children did not thrive. They often began to do poorly in school and were more likely, as we found in a study of approximately eight hundred children from more than three hundred families, to suffer from parasites such as ascarids, which can impede development.

## Visual Art and Propaganda

In order to influence behavior in particular directions, humans regularly used visual and other arts. Miller, who refers to this as the propaganda theory of art, dismisses the idea. He claims that in small, prehistoric bands, there would have been no one with the incentive to spend the time and energy producing group propaganda. Language is surely a much more efficient tool for telling people what to do and what not to do. Only large institutions that can pay propagandists usually produce propaganda (263).

While Miller has made an interesting proposition, it is not clear how he defines propaganda. *Propaganda,* according to its dictionary definition, refers to information deliberately spread in order to influence behavior. As public health officials learned long ago (probably tens of thousands of years after mothers learned it), the spoken word alone can be remarkably ineffective in influencing behavior change. Influencing behavior is much more complex than just telling someone what to do or giving them the information they need to make an "educated choice" (Coe and Keller 1995). Mothers would have had a great deal of interest in spending time and energy producing propaganda, not for some ill-defined group but to influence offspring and descendants. Although Miller dismisses the propaganda theory of art, he recognizes that art can influence behavior. Its aim, after all, is to seduce.

## Kinship and Descent

I have already demonstrated the impressive amount of visual art that is related to kinship, descent, and ancestry, and I have shown how visual art identifies ones' kin, ancestors, and codescendants. Miller, however, follows Amotz Zahavi in mocking the species recognition idea as attributing a very high degree of stupidity and very poor mate choice to animals (203). Do we, he wittily asks, spend hours in courtship so that we can distinguish that the other individual is a human rather than a chimpanzee?

There is no trouble distinguishing between chimpanzees and humans because there is a huge gap between the two. However, such a task may have been more difficult in other periods of evolutionary history; many potential intermediate lineages have apparently died out. Consider Europe 70,000 years ago, or most of the past four million years in Africa, where numerous lineages of hominids coexisted. Consider the importance of identifying species and even subspecies among tropical birds when many closely related species live in the same area. A poor mate choice is costly for females. Anything that helps a female (or her kin) make a good choice is important.

A species is made up of individuals who share descent from a common ancestor. To the extent that individuals inherit their decoration from their ancestors, they necessarily communicate their ancestry, whether or not they communicate that they are extremely fit because they can afford costly ornaments. In the case of animal decoration, it may be the permanent sameness (the elaborate tail of peacocks) that advertises ancestry, not the unique differences in such things as intensity of coloration, which may signal something else. We do not always know what such decoration is advertising.

In the three-spined stickleback males (*Gasterosteus aculeatus*), the red throat coloration varies among males, and the intensity of the coloration is not always correlated with physical condition (Bakker and Mundwiler 1994; Frischknecht 1993). Males with a bright red throat may have "reduced mating probabilities" in the wild because the red attracts predators. A brighter male also may experience increased risk of injury and death from conspecific males (Kraak, Bakker, Mundwiler 1999, 703). As male sticklebacks care for their young, color intensity may be inversely correlated with offspring survival. It is not yet clear precisely what the benefit associated with having a brighter throat than a fellow codescendant's might be.

Humans have regularly used tribal and clan dress and ornaments to identify kin and codescendants. In a tribe, some individuals share a closer ancestor (the clan ancestor), while members of all the clans share a more distant ancestor (the tribal ancestor). While humans presumably can identify close kin because they interact with them regularly, body decoration can identify both one's clan and one's tribal ancestors and codescendants. In other words, it can distinguish closer relatives from more distant relatives. The degree of cooperation is apparently correlated with the closeness of kinship ties. Cooperation with family members was more intense than was cooperation with clan members; cooperation with clan members was more intense than was cooperation with tribal members (Palmer and Steadman 1997). Cooperation with those outside one's tribe was limited. When females (and their kin) use such decoration to select a mate, they are, given inheritance of designs, using body decoration to identify a male's ancestry and traditions as well as the number of his kin. While human body decoration undoubtedly differs from what occurs in animals, it may be important for all organisms to be precise in identifying mating distance not only between species but also within a species.

## Morals and Marriage

While I agree with Miller that moral behaviors can have high costs, I do not agree that this means moral behaviors are necessarily fitness enhancing. It is easy to think of examples of behaviors considered moral that would not be fitness indicators. Becoming celibate or giving up all one's worldly goods have high costs and are referred to as moral behaviors, but they are not fitness enhancing.

Moral rules that have high costs are frequently depicted in the visual arts. One universal behavior, marriage, is encouraged by moral rules and has been depicted in and encouraged by traditional visual art and its associated songs and chants (Cory 1956). Yet marriage reduces the total number of mates a male can acquire and the total number of offspring he can sire. Marriage means that not all women are open game. To resolve this problem, Miller argues that marriage occurred late in human evolution:

> Most children were probably born to couples who stayed together only a few years. Exclusive lifelong monogamy was practically unknown. The more standard pattern would have been "serial monogamy," a series of nearly exclusive sexual partnerships that were socially recognized and jealously defended. Relationships may have sometimes ended amicably, but perhaps more often one partner would reject or abandon the other, or one would happen to die. This is the pattern of most human hunter-gatherers, because they do not have the religious, legal, and property ties that reinforce ultra-long-term monogamous marriages in civilized societies. (186–87)

I argue that marriage is ancient and does not require a legal system or property ties. Monogamy, not marriage, is a recent human cultural behavior.

"Immoral acts," Miller (2000) writes, "are mainly those we would be embarrassed by if our boyfriend or girlfriend found out about them" (307). What he ignores here is how we would feel if our parents found out. For most of human history, children have learned moral behavior by listening to and watching their mothers and fathers. As immoral behavior would, by definition, involve a rejection of that teaching, parents, mothers in particular, would be the first ones we would worry about offending. When traditions break down, we reject ancestral/parental teachings. However, if we are embarrassed because of what our girlfriend or boyfriend may think, it is presumably because she or he also accepts the same (ancestral) propaganda.

## Making Sense of Magic, Religion, and Supernatural Claims

Anthropologists, Miller writes, "have suggested that the principal function of art during human evolution was to appease gods and dead ancestors and put people in touch with animal spirits" (2000, 263). While Miller appears to think that anthropologists were duped (as there is no evidence that anyone actually was in touch with the supernatural), anthropologists could hardly have ignored the supernatural talk that regularly was used to explain visual art. Supernatural talk is ubiquitous in the societies anthropologists study. Further, anthropologists are not the only ones who have been led to explain visual art by reference to belief. Arnold Hauser insisted in *The Social History of Art* that "any other explanation of Paleolithic art . . . is untenable" (1959, 1:9). According to Miller, people who believe that "gods, ancestral ghosts, and animal spirits" really exist are "deluded" (264). With one clever phrase he has managed to sweep under the rug the majority of humans who now live or have ever lived. It may be naive, however, to assume that these people are deluded, just as it may be naive to think that what people say is an accurate reflection of what they think. People might tell us what they know we want to hear; they may exaggerate or lie, have an inadequate vocabulary, or be confused, unsure what motivations lie behind their behavior (Anderson 1979). Because Miller assumes that people are rational (and rational means that they will always do what makes sexual or reproductive or economic sense), he cannot understand why humans claim, for example, that a sandpainting can cure an illness.

Religion, Miller insists, has survival value only if it "actually does what is claimed" (264). If the alleged spirits do not exist (a fact that cannot be proved or disproved), "there is no survival or reproductive advantage to be gained from appeasing or contacting them" (264). This "delusion," Miller writes, "is not evolutionarily stable" (264). The only way such behavior makes sense, he argues, is if "others grant the religiously imaginative individual higher status or reproductive opportunities" (264). As religions tend to be conservative, it is not clear how many individuals would have behaved imaginatively. However, for Miller, sexual selection and competition answer our questions of religion, even though religions are often traditional and conservative and religious leaders teach and are expected to practice self-restraint.

Miller draws on his psychology background to design a double-blind clinical study to see if a Navajo sandpainting ceremony (in which ancestral spirits are evoked to heal the sick) can be proved to be medically effective. Such a study might be difficult to design. What constitutes a cure? Further, if the ritual were to have no curative effect, all chanters would have to say is that the ancestors' spirits had failed to show up that day because so many psychologists were present or that evil spirits had shown up and pretended to be good spirits. The important point here is that it may be irrelevant whether or not the ritual works.

A direct relationship between a supernatural claim and an outcome may not be important to some humans. People apparently continue to pray even if their prayers are not immediately or, for that matter, ever answered. They continue to perform rituals even if the rituals seldom or never bring about the hoped-for outcome. While science may be built on a measurable correlation between an action and a result, religion clearly is not.

Isn't there a better explanation for religion, one that does not require millions of people to be regarded as fools yet is compatible with natural selection? In ancestor worship, the claim is regularly made that ancestors are influential. They are concerned about, and in many cases are said to interact with, their descendants, involving themselves in the lives of the living. In his book on Late Shang China (ca. 1200–1045 B.C.E.), Keightley writes, "The cult of the ancestors enlarge the dimensions of the human community to include the dead, as the powerful members of the lineage who had flourished in time past continued to influence and demand the attention of their descendants who lived in time present" (2000, 101).

As it cannot be shown that ancestors care about and involve themselves with the living, what can be done with this claim? It is precisely on the word *claim* that Steadman, Palmer, and Tilley (1996) focus; it is this claim, they argue, that is universal. Supernatural claims are regularly, indeed universally, made and readily accepted. Yet there is little interest among traditional people in proving supernatural claims. No one asks, for example, how we know that ancestors are involved with the living, rolling over in their graves because they are unhappy about something we did. The way to study supernatural claims empirically, Palmer and Steadman (1995) argue, is to ask what the identifiable effect of the claim is on the living. Acceptance of such a claim, they argue, promotes cooperation. The persistence of traditions, the ferocity of the fighting force of an ethnic group mobilized to right wrongs committed against their ancestors and codescendants, and the sacrifices made in so doing, all attest to the cooperation associated with ancestral supernatural claims.

It is hard to escape the fact that humans seldom form enduring social relationships with individuals who treat all of their statements with skepticism. If, on the other hand, a supernatural claim is made (for example, a ghost lives in the old abandoned house) and others accept it on faith and begin to act as if the claim were true, it is hard to argue that this does not involve a form of cooperation. While religion clearly may involve much more than this, supernatural claims do seem to set up conditions that promote cooperation.

For Miller, art involves one individual competing against all others, even those who are kin. This assumption makes it difficult to account for art that is widespread, namely the tribal and traditional art described in this book as being characterized by the promotion of cooperation among a category of individuals claiming common descent. Although I would be foolish to deny the competitive nature of art, what seems to occur is that codescendants of one ancestor compete with codescendants of another. What may be unique about some human religions is the appearance of a prophet who creates a kinship category out of members of disparate tribes.

Miller and I do agree on one fundamental point: An investment made in courtship is at the expense of an investment made in survival. I would argue, however, that human ancestors regularly encouraged their descendants to make a large investment in parenting and in codescendants and to restrain their courtship behaviors. One has to ask whether or not the tremendous investment that humans made in being social (K-behaviors) was not at the expense of both their survival and their reproduction.

Miller, by ignoring anthropology and parenting behaviors, leaves out evidence that seems to be crucial in testing his ideas about culture. He is not alone in ignoring traditions. Unless we can account for traditions, we cannot assume that we are able to account for much of human behavior. In accounting for traditions, however, we cannot dismiss the majority of humans who have lived as having been duped. Inherited traits that endure—traditions—must have a function.

# CHAPTER 8

# Testing the Ancestress Hypothesis

When ancestors are lost, competition intensifies internally. Visual art becomes more competitive and a runaway type selection occurs. When creativity is valued, elders become obsolete and their conservative ways are discarded. Visual art is characterized by a focus on individuals, creativity, change, notable demonstration of dominance, and the breakdown of any traditional rules of art, including, for example, those of balance, symmetry, and color combinations. The ancestress hypothesis, thus, is not falsified by evidence that visual art is competitive and creative and no longer has ties to ancestry. It predicts this should occur under certain circumstances.

## How to Test the Ancestress Hypothesis

There are at least four tests to identify conditions and situations that might threaten the ancestress hypothesis:

1. Early prehistoric art or traditional visual art that had an explicitly sexual content or that appears to have been made by males, especially for sexual stimulation, would threaten this hypothesis because it would promote competition between male kin and threaten female choice.
2. If no limitations are placed on the visual art produced or used by a dominant male in a traditional people, my hypothesis will be weakened. I predict that among traditional people limitations will be placed on male decoration and a male's ability to use art to attract attention to himself as a unique individual. When dominant males begin to decorate themselves splendidly, they will justify the decoration by reference to ancestors. As direct descendants of a particular god or ancestor, for example, they inherited the right, and even the duty, to use the decoration.
3. My hypothesis is threatened if traditional people readily adopt a new visual art style from nonkin, particularly if they use the new motifs on ritual objects. If new styles are adopted, they will be adopted from kin or metaphorical kin and/or the claim will be made that changes came from or are approved by the ancestors.
4. When ancestors are important, visual art is conservative and used to promote cooperative social behaviors among kin and codescendants. When different cultures come together, however, and change begins to occur, these changes will be based either on metaphorical ancestry

(and art will promote cooperation among the new kin) or they will be at the cost of ancestors and related to a breakdown in kinship and descent systems and an intentional and often antagonistic discarding of traditions. This last prediction has a number of implications, two of which are that traditional art will not be disrespectful of elders or ancestors and that nontraditional art will criticize the past either explicitly, by using visual art to criticize ancestors and traditions, or implicitly, by ignoring ancestors and their particular visual art styles.

## Prehistoric Art Is Sexual, Not Maternal

Miller predicts that males are the originators of prehistoric art and proposes that evidence supporting this will be "a lot of sexual content" in the art (274). Early art scholars would agree with him: The ancient carving of cupules in stone have been referred to as female vulvas carved for males, and the so-called Venus figurines have been explained as sexual icons. I predict that visual art in early prehistoric sites, as in traditional people, would be conservative and show evidence of an association with maternal concerns, including kinship and descent relationships. My hypothesis is weakened if visual art found in prehistoric sites and among traditional people has explicit sexual content (for example, depicting various positions for sexual intercourse). Such art, if found among traditional people, will be made for secret male rituals, or idiosyncratic, made, for example, without parental approval by teenage boys trying to sexually arouse themselves, as seems to be the case among the Dani.

As we will never have the complete story on the social origin or early usage of visual art, this hypothesis is difficult to test. Early artifacts may have been made of perishable material, may have washed out to sea, or may have been buried under piles of rocks or layers of sand. Further, it is impossible to know with certainty who made the art, a male or a female, and for whom it was made. It does seem to be true, however, that the sexual division of labor, which is universal, implies that males and females have different activities. Pottery making and weaving, according to the ethnographic record, tend to be female activities.

As the best inference is from the known to the unknown, there is reason to suspect that females made these early objects. Based upon this logic, scholars have argued that by the Upper Paleolithic women's cultural activities (weaving baskets and textiles and making clay figurines) had reached a level of technological sophistication that suggested "significant antecedent development" (Soffer, Adovasio, and Hyland 2000). These cultural activities were not only widespread but also showed temporal continuity (Davidson 1997). Given this persistence, I argue that mothers, not just females, played an important role in the production of this early visual art. Humans apparently have long lived in small groups of kin. The transmission of these skills, thus, would have been between generations of kin, probably mother to daughter.

## Red Ochre

Humans have used red ochre for a long time, perhaps longer than 300,000 years; shaped red ochre crayons were possibly being made as early as the Middle Paleolithic (Deacon 1989). From about 25,000 B.C.E. onward, there has been heavy use of red ochre; in some caves in Western Europe, particularly France, cave floors dating back to the Aurignacian were impregnated with reds to a depth of eight inches (Leroi-Gourhan 1968). Red ochre also is frequently found in burials (Wreschner 1980). Humans apparently have been designed to respond to red, distinguishing the electromagnetic waves at the red end of the spectrum most easily and infallibly. Males, however, are much more likely to have red-green colorblindness than are females (see Knight 1991; Barber 1991). Females, thus, may have been the first to notice and begin using the color red (see Knight 1991).

## Cupules

Cupules, or cup-shaped hollows carved in stone that vary in size and depth, are said to be among the oldest art objects. They are found in nearly all parts of the world (Grieder 1982), often in combination with other forms such as circles and grooves. They often are found in mountains or caves or in sites once located near water. At La Ferassie, a large stone laid over a child's grave had a cupule ground into the underside (Sanders 1968, 7). In Kimberley, Australia, panels of cupules have been found. In one rock shelter, the vertical sidewall of one rock has more than three thousand cupules pounded into the surface. Each is about the size of half a tennis ball. While scholars do not all agree, Gambutis (1991) argues that the earliest cupules date back to the Middle Paleolithic (100,000 B.C.E.) and that the practice of carving cupules persisted through the Mousterian (40,000 B.C.E.). Cupules are also found in later sites; there are cup marks on menhirs and megalithic graves in Western Europe (Gambutis 1991).

While there is little ethnographic evidence available that might help explain these cupules, many archaeologists and art historians have claimed that they represent vulvas made by males (see Grieder 1982 for a discussion). However, the small amount of ethnographic evidence that exists seems to paint a slightly different picture. Garrick Mallery (1893, 196) claimed that women in India still used what he called cup petroglyphs. "At this very day," he wrote, "one may see the Hindu women carrying the water of the Ganges all the way to the mountains of the Punjab, to pour into the cupules and thus obtain from the divinity the boon of motherhood earnestly desired." In the 1960s, the Pomo of California pecked cup marks onto boulders when they wished to conceive a child (Heizer and Nissen 1977).

## Venus Figurines

Many of the figurines recovered from prehistoric sites in Europe initially were referred to as Venus figurines. Because they were often, but not always, female, archaeologists assumed that they had been made by males and served some sort of erotic function. These figurines have been found in a geographic range stretching from the Pyrenees to European Russia, with figurines known from southwestern France, southern German, northern Italy, Czechoslovakia, and the Ukraine (Davidson 1997).

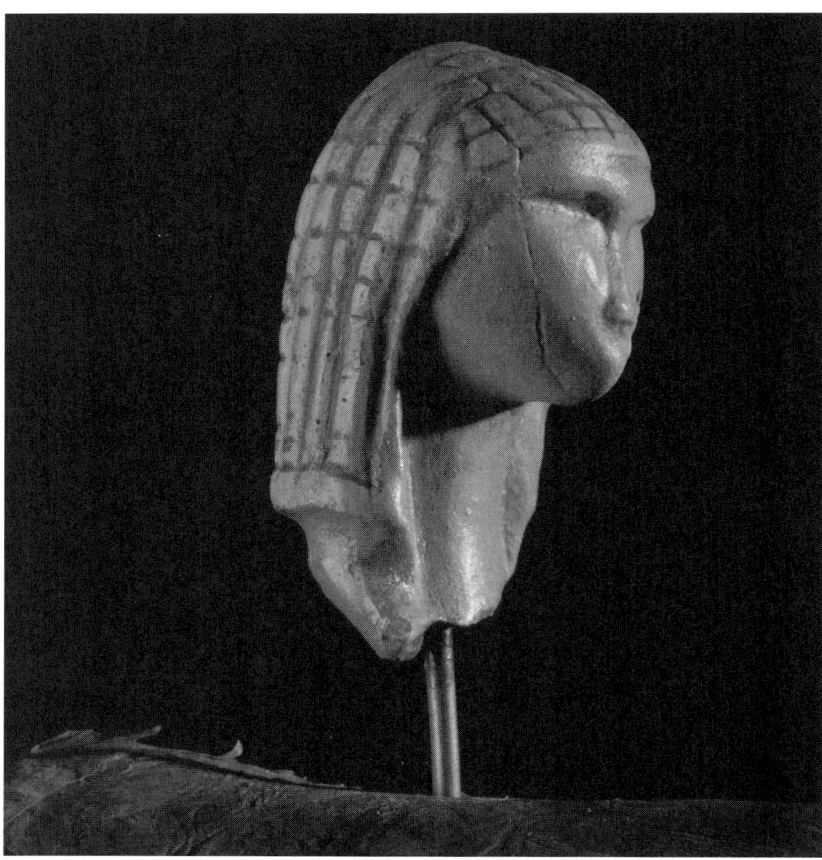

**Fig. 32.** The Lady of Brassempouy (ivory or mammoth bone), from Landes, France, circa 21,000 B.C.E.). Photograph by Erich Lessing/Art Resource. Courtesy of the Musee des Antiquites Nationales, St-Germain-en-Laye, France.

The figurines are quite old. While many of them cannot be dated, the oldest may be a female figure carved in volcanic tuff that dates back somewhere between 233,000 and 800,000 years ago. This figurine was found in the Acheulian site of Berekhat Ram in the Near East (Marshack 1996). The figurines from Central Europe seem to be dated about 28,000–24,000 B.C.E., while those from the Russian plains are somewhat later than this date (see Davidson 1997). Despite their antiquity, many are beautifully carved, sometimes in stone, sometimes in bone or ivory, sometimes in coal. The figurines are small and portable (some may be pendants) and are often of women. They range from explicit and realistic to highly stylized carvings and even dubious human shapes (Delporte 1979). Female images were also carved in bas-relief and engraved on stone and bone.

These figurines show continuity over thousands of years. There is some evidence that there was a shift of people, who carried figurines with them, from Central Europe to the Russian plains (Soffer 1993). Similar figurines later appear in the Neolithic (Nelson 1997) in various parts of the world.

According to Barber (1994), at least two of the figurines are depicted wearing a sort of garment known as a string skirt, which consists of a belt band from which numerous strings hang down to form a skirt or apron. The Venus from Lespugue in the Pyrenees wears her skirt only in the back, but

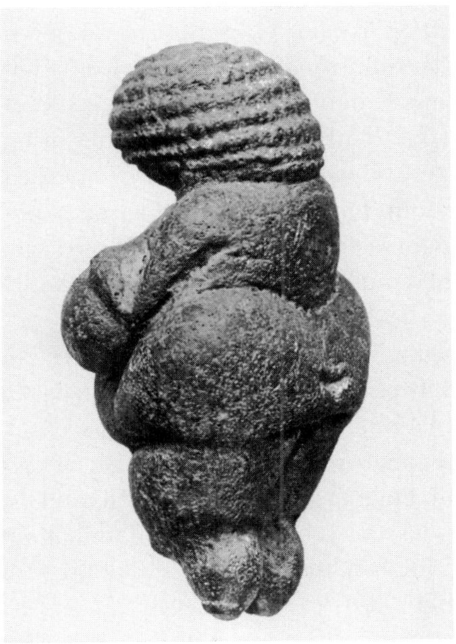

**Fig. 33.** Rear view of Venus of Willendorf (limestone), Upper Paleolithic. Courtesy of the Naturhistorisches Museum, Vienna, Austria and Giraudon/Art Resource, New York.

on each string the carver has clearly shown with little incisions the twists of the cord and indicates that the bottom of each twisted string frays out into a mass of loose fingers. Similar skirts have been found in Denmark in Bronze Age sites. Some figurines have other decoration that does not lend itself so readily to interpretation. While it is not clear if these designs portray ornamentation, clothing, tattoos, or scarification, later figurines from southern Greece have similar markings that resemble the tribal markings illustrated in ethnographic reports.

Excavations of Paleolithic settlements in France, Eastern Europe, and Siberia may suggest that females used these figurines. The characteristic position in which the dolls or figurines are found in an excavation is inside the dwelling hollow, close to the hearth, and frequently in storage pits dug into the floor (Gambutis 1991). The figurine from the Abri Facteur was found near the wall of a shelter, in an area marked by a paucity of stone tools (Delporte 1968). At Dolni Vestonice in Moravia, the most complete figurine was associated with a large hearth (Klima 1957). A figurine from Kostienki, dating to about 23,000 years ago, was upright in a small pit, leaning against the wall and facing the center of the living area and the hearth (Bahn and Vertut 1988).

The Soviet archaeologist Abramova (1967) associates the figurines with the "spirit of the hearth," as there is a connection between the figurines and the traditions of the Tungus and other northern Asiatic people. These people continue to keep in their tents a portrayal of a hearth spirit visualized as a clever, old but still strong and vigorous woman. Patricia Rice (1981) suggests that these female figurines may represent several stages of womanhood. She asked respondents to rate the figurines in regard to age (juvenile, reproductive age, and old), and whether or not the figurines appeared to be pregnant. She found a high degree of agreement among the coders, suggesting that the figures do represent variability in age. This suggests, she writes, that the figures could have been used to teach girls about the female life cycle.

The decoration evident on some of the figurines could represent hair arrangements and permanent body markings such as tattoos and cicatrization. In living populations, these body markings often indicate that the wearer has passed through some rite of passage, often moving from girlhood to womanhood. Further, these markings are ancestral designs associated with particular descent groups. Anyone familiar with specific designs within a region can identify the wearer's affiliation with a particular clan (Seeger 1973; Hambly 1925).

Less than a century ago, figurines found in many parts of Africa were used at ceremonies celebrated when a girl was physically and emotionally developed enough to assume adult activities and responsibilities (Cory

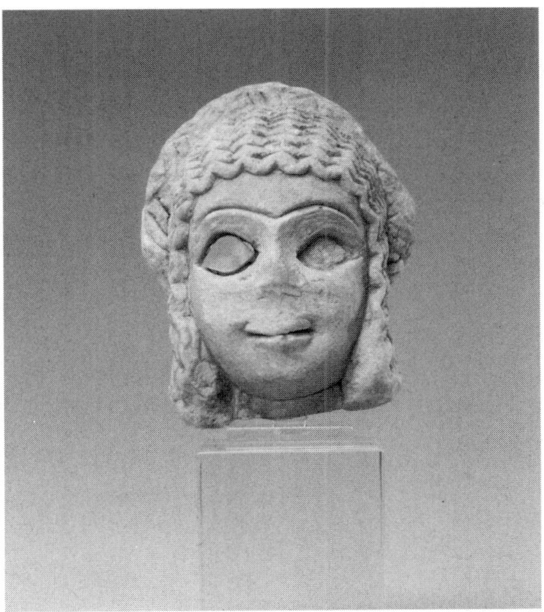

**Fig. 34.** Female head (gypsum) probably from Diyala Valley (Klalaje?), circa 2700–2500 B.C.E. Courtesy of the George Ortiz Collection, Switzerland.

1956). The figurines were used to teach girls about marriage, sexual matters, and their culture's value system. Before the rituals began, kinswomen were responsible for making the figurines, which were often made of clay and always had stereotyped faces and rigid forms. The figurines were referred to as sacred because the model for them came from the past, inherited by all the "descendants of the same ancestor" (18–19).

During the puberty rites, figurines were placed in front of the novices, a song connected with the figurine was sung, and the purpose of the figurine was explained. These ritual figurines were associated with moral lessons that taught the importance of passion and the advantages of restraint; they were "intended to instill a sense of discipline into the novices" (29). These "figurines not only act as mnemonics for the songs, but also come to present moral attitudes and obligations and to represent them in quotable form . . . [that] the girl will remember" (163). Figurines also were used to transmit tribal wisdom related to birth and death.

Cory described how the figurines, and the sayings and songs that accompanied them, taught girls to respect their elders and the traditions of their ancestors: to marry and look after her husband properly and to be a good mother. One figurine was of a hornbill. The story associated with it taught girls that the hornbill "builds no nest, but lays her eggs in a hollow tree. If

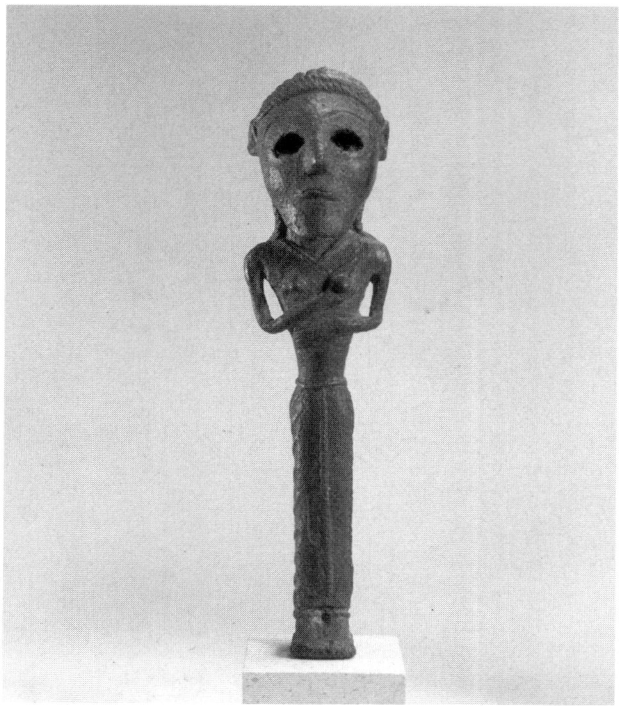

**Fig. 35.** Female figure (copper) allegedly from Syria, though it is more likely from central-south Lebanon, circa 2000 B.C.E. Courtesy of the George Ortiz Collection, Switzerland.

you do not look after your husband properly he will fly about like the horn-bill without building a nest" (59). A female figurine with breasts was used to council girls to "marry while you are young. Do not stay in your father's house forever. The years go by quickly and suitors will disappear when breasts lose their firmness" (61). A clay rhinoceros was used to discourage novelty and promote the retention of traditions. The story told about the rhinoceros was that "the rhinoceros is proud of his horn, but it may cause him to die of thirst" (53). According to the rhinoceros song, all other animals have their horns branching sideways from their heads; but the rhino has his on his nose. If water becomes scarce in the dry season and only shallow pools remain, the rhino may die of thirst because his horns keep him from drinking in shallow pools. This taught girls that it is not good to try to change old traditions. There is nothing to be proud of in an invention that may kill the inventor. Follow the teaching of your mother and the old women and do not start trying out new customs (53–54).

In South Africa, a South Sotho girl, just before her marriage, was given a

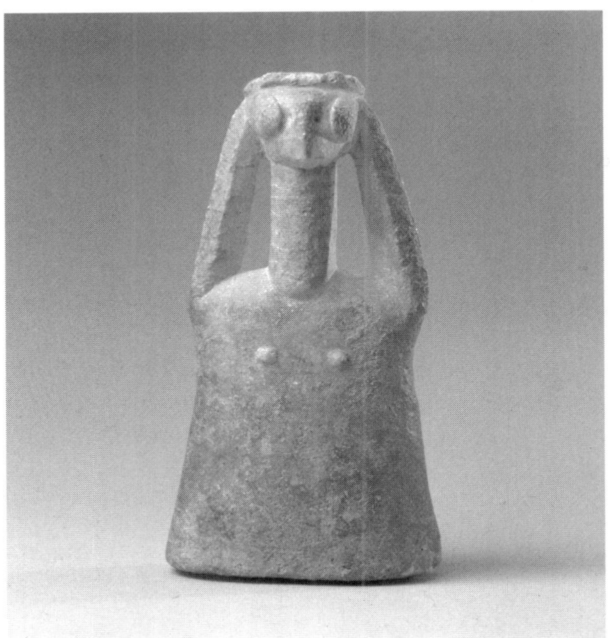

**Fig. 36.** Group of Anatolian figures, including female mourner (alabaster), Anatolian from Asia Minor, early to middle Bronze Age. Courtesy of the George Ortiz Collection, Switzerland.

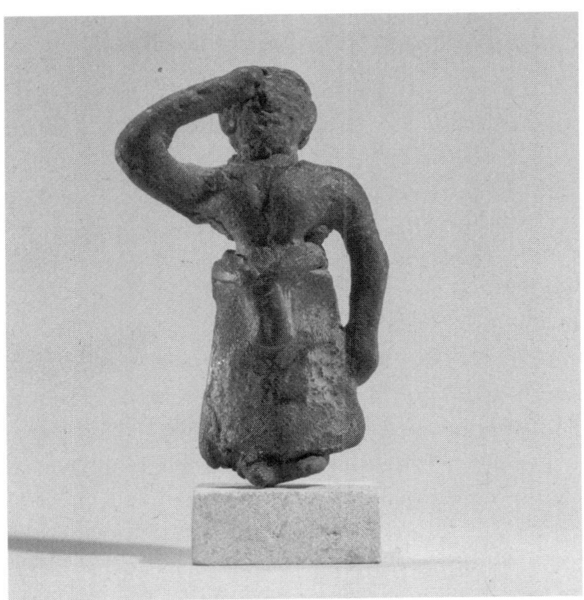

**Fig. 37.** Female votary (bronze) from Greece, circa 1700–1580 B.C.E. Courtesy of the George Ortiz Collection, Switzerland.

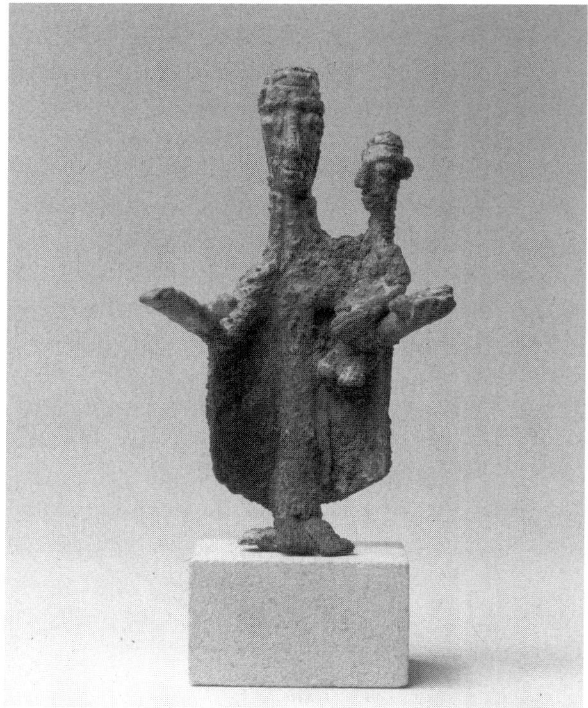

**Fig. 38.** Mother and child figure (bronze) from Lode (region of Siniscola), circa tenth to ninth B.C.E. Courtesy of the George Ortiz Collection, Switzerland.

small clay bride doll made for her by her grandmother. The gift of the doll meant that the girl was socially and physically ready for marriage. On her wedding day the girl carried the doll to show that she wished to present her husband with children. The doll was kept by the bride as her most precious possession until "such time as she is blessed with a child" (De Lange 1961, 94).

Among the Bemba, initiation images were modeled by the mistress of ceremonies (usually a midwife). Figurines often were used to illustrate narratives or proverbs extolling the virtues of hard work, courage, generosity, honesty, and the importance of the family unit. Bantu figures could be stored in a corner of the initiation hut during the lengthy ceremonies and covered with a cloth; eventually, they were buried with the afterbirth of an initiate's first child. Until recently, a pregnant Igbo woman of eastern Nigeria carried around with her a small bag containing a wooden doll (Ucko 1968). Zuni Indian women of the southwestern United States traditionally carried magic dolls following a miscarriage or when they wanted to get pregnant (Parsons 1919).

In sum, although there is no direct evidence proving that these female figurines were made by our ancestresses, there is ample ethnographic, associative, and anecdotal evidence supporting that this was a possibility. The fact that dolls were used to educate males does not disprove this proposal, as the dolls were used to educate males about proper social behavior and self-restraint (Cory 1956). In the literature, I was able to find one account of a female figure carved of wood that was clearly associated with males. According to Margaret Mead, the clitoris of this figurine was painted bright red. The carving was used in the secret ceremonies held in men's houses in New Guinea. Across cultures, however, females were by and large the ones who used dolls.

## Competition: There Are Conspicuous Differences between Individuals

The sexual selection hypothesis, and the assumption that individuals behave in ways to promote their own best interest, leads to predictions that there will be conspicuous differences among individuals. In Mount Hagen, New Guinea, there is some variation in male body decoration. Strathern and Strathern write that although men dress "similarly [they] should be distinguishable from each other" (1971, 99). The word *similarly* here is important. Among many people, calling undue attention to oneself by showing arrogance or by boasting is discouraged. These characteristics, Lee writes, "are strongly disapproved of by the !Kung; humility is the proper stance and the society has devised an array of humility-enforcing devices for bringing people into line" (1979, xx).

Heider explains that "although there are more and less important men in Dani society, the fact is known to all so that it does not need to be communicated through attire. A stranger looking only at attire could not tell which of a group of men is the most important and which is the least" (1979, 56). The Big Men of the Dani in New Guinea were men who had more status and often more responsibility, although they did not necessarily have more influence or wealth. All Dani males, however, not just Big Men, wore penis gourds that varied in length from navel height to chin height, some straight, others curled at the ends, some plain, others festooned with furry marsupial tails sticking out at the tips. Heider found "no correlation between the gourd of the day and either long-term and short-term personality" or differential status (56). Heider claimed that he could find no function for the gourd, except it concealed (but did not protect) the penis and it was awkward to wear. When cotton trousers were introduced, however, most Dani continued to wear the gourd, not the more comfortable pants. The practice persisted despite the fact it was uncomfortable and had,

Heider writes, "no meaning and little function" (57). One function that the gourd did have, of course, was that it communicated that all these men were Dani.

In Australia, all Mardujara males, Tonkinson (1978) wrote, wore the same clothing. This clothing consisted of a thin hair-string belt used mainly for holding small game to free the hunter's hands (10). No male attempted to discover other ways to decorate himself more profusely or elaborately.

Masaii warriors were expected to wear elaborate headdresses during wars and raids. Klumpp wrote that the warriors "are expected to spend a great deal of time and attention tending to their appearance, primping and swaggering. They spend many hours in small groups doing each other's hair and organizing their ornament assemblages" (1987, 172). While this decoration of young warrior males is interesting, the process of decoration was social, not solitary and not necessarily competitive. It is also of interest that "at the end of junior warriorhood and again at the end of senior warriorhood, men relinquish ornaments and start over again with progressively reduced assemblages" (160). In other words, as a male reaches the most prestigious stage of warriorhood, he is less elaborately dressed.

The Yoruba, prior to extensive contact with colonial government, had sixteen sacred crowns, one for each of the ancestors of the city-states. These crowns were said to be so sacred that they could not be worn, although Yoruba rulers, Blier writes, made and "sometimes wore crowns of these types" (1998, 79–80). During the nineteenth and twentieth centuries, when Yoruba royal authority was much diminished due to colonial presence, the number of Yoruba rulers wearing crowns increased even though the crowns no longer communicated legitimate influence. Although the crowns were used more often, their usage "closely coincided with a general loss of traditional authority" (80). What the crown continued to do, however, and to do so more often and more widely, was to communicate ancestry.

Perhaps the strongest evidence of a link between kinship, ancestry, and visual art is found in the facial tattoos of Maori males and females. While the tattoos of the females were quite simple (confined to the area near the mouth) compared to those of males, the tattoos of both males and females followed a particular format. Female tattoos, Simmons (1986, 105) explains, communicate descent. They also could have communicated whether or not the female was legitimate and, if she was illegitimate, whether or not the father acknowledged paternity.

Sydney Parkinson, artist with Captain Cook on his first voyage in 1768–71, wrote that on the south side of the Tunarganui inlet in New Zealand, the males were "tattooed in various forms on their faces" (1784, 89). On the other side of the inlet only the lips were tattooed. The males on the north

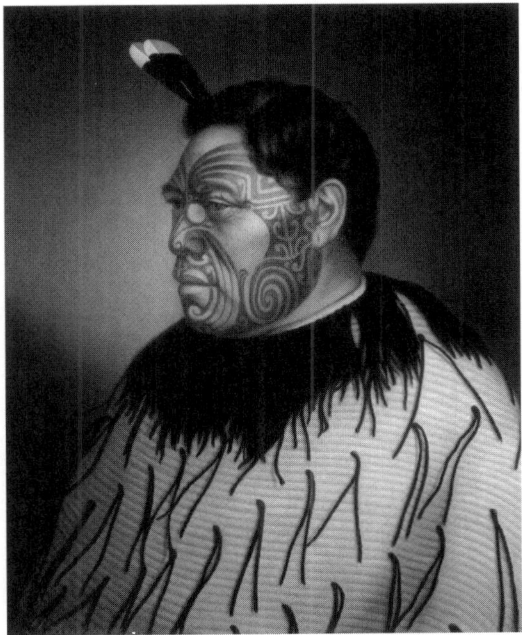

**Fig. 39.** Male tattoos of Tamati Waka Te Puhi, Maori of New Zealand, painted by Gottfried Lindauer, circa 1876.

side of the inlet never attempted to increase the size or number of the tattoos they received so that they could outshine the males on the south side. No male, Parkinson wrote, attempted to obtain tattoos that distinguished him as a unique individual. Males used the tattoos to which they were entitled. While the tattooing process was painful, and sick or weak males may not have been able to endure it, in generation after generation, males continued to get tattoos and those tattoos apparently remained confined to the face.

In regard to Maori male facial tattoos, Simmons (1986, 131) writes that the right-hand side of the face conveyed information about the father's tribal affiliation. The left-hand side of the face conveyed information about the mother's tribal affiliation. The forehead area was reserved for indications of rank, which were often tied to lineage. The area around the nose indicated tribal affiliation. The area from the temple to the ear contained information about the father's or mother's lineage, descent (for example, whether from first or second marriage), and rank or birth order. In the area between the midcheek and the jawline, it is communicated what the individual does (often inherited jobs such as gardener or carver). Although

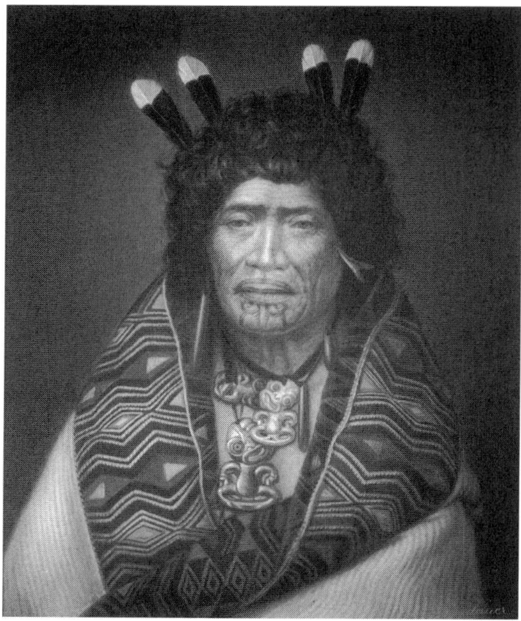

**Fig. 40.** Female tattoos of Rangi Topeora, Maori of New Zealand, painted by Gottfried Lindauer, circa 1876.

brothers may both have identical maternal and paternal tattoos, one brother will be older than the other, and that will be communicated in the tattoo (Mead and Kernot 1983). While the decoration of the brothers is similar, the two tattooed faces are distinct. As the majority of the designs generally communicate ascribed, not achieved, status, few tattoos communicate a male's individual prowess.

These same tattoo designs were often carved into wooden figures. Ritual objects and the center posts of houses displayed relevant tattoos (Simmons 1986, 146). Later, tattoos began to be used as signatures. The signature and the face both identified a specific individual and that individual's entire ancestry.

Some might argue that the leaders of human hierarchies used ancestral visual art to increase their own personal prestige and power. There appears to be some truth to this, once humans adopted male hierarchies with differential access to resources. Chinese commoners, for example, were not allowed to worship ancestors beyond the generation of their grandparents. Among the Maya of Mayapán, each lineage was ranked by the antiquity of its pedigree. According to many scholars, "the descent line of the noble families is deeper and more venerable than that of the commoners" (McAnany 1995, 116).

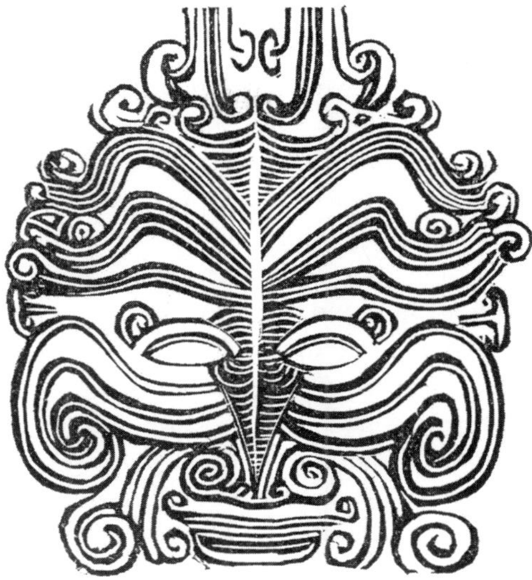

**Fig. 41.** Signature of Kowiti, a chief of Waimate and Maunganui, Maori of New Zealand. Courtesy of the Auckland War Museum, Auckland.

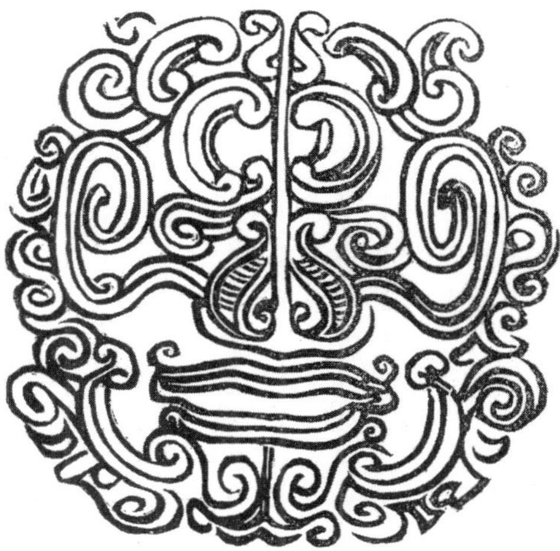

**Fig. 42.** Signature of E'Gnogni, a chief of Mukou, Maori of New Zealand. Courtesy of the Auckland War Museum, Auckland.

A point to consider here is that, even in a hierarchical society, the more distant the ancestor, given reproductive success, the greater the number of codescendants that can be identified. By limiting individuals to only three generations of identifiable ancestry (child, parent, grandparent), it is possible to significantly limit the number of individuals that any person can identify as codescendants. It thus limits the number of individuals that a person can identify as kin and, thus, the number of individuals who are likely to behave as if they were close kin and offer assistance when necessary.

Alexander (1979) argued that individuals copy the body decoration of those who are famous and influential, even if that copying involves considerable discomfort. It is true that in the prehistoric and historic records we see significant evidence of copying. The fact that sumptuary laws (or laws prohibiting such copying) have existed makes it clear that humans have copied those of higher status and that those of higher status, who presumably made the laws, did not want the copying to occur. Given strict marriage rules, however, sumptuary laws kept others—outsiders—from usurping the decoration to which true codescendants were entitled. Further, copying a dominant male cannot explain traditions. Copying dominant males would lead to the prediction that change would occur regularly, with each generation copying the new, dominant male. Each generation should produce its own dominant and influential males. If decoration is copied through time, as traditions indicate, then humans must have been copying their ancestors.

## Adoption of Visual Art Style from Nonkin

The ancestress hypothesis predicts that a new style of visual art would not be readily adopted from nonkin. If visual art styles are adopted from nonkin, then these changes would be justified by reference to kinship and ancestral approval. If humans readily adopt art from nonkin or without using a kinship metaphor, the ancestress hypothesis would be weakened.

Buddhism arrived in China around the first century, bringing with it the image of Buddha (Wu 1989). Buddha was always seated on a dais in a rigid frontal pose. After the introduction of Buddhism, the Chinese deity, the Queen Mother of the West, began to be depicted in a new way on tomb murals. While she had always been shown in profile, she was now seated, like Buddha, and had a frontal posture. As Buddhism took deeper roots in China, the frontal posture began to depict all religious deities. In this example, the old ancestors (the Chinese deities) are merged with the new metaphorical one (Buddha).

When Mexican artists, after the Conquest, adopted European art styles, they used the new techniques to paint portraits of the new, metaphorical ancestors, the Christian religious figures. After the Mexican Revolution (1910–1920), José Vasconcelos, secretary of public education, argued that the Mexican people were a cosmic race, made up of the offspring of Spanish fathers and Indian mothers (Riding 1985, 201). Art students and intellectuals began to attack the Eurocentricity of Mexican visual art. A focus of their criticism was the academy, the neocolonial institution where European-trained artists, at least until the time of the revolution, taught courses in Western-oriented fine art. The new art that emerged, out of an effort to bring the country together after the revolution, was drawn from the ancestors of the living Mexican people, ancestors who largely had been forgotten. Ancestors were seen as the repositories of Mexican national identity and the starting point for what was called a "new aesthetic" (Bakewell 1995, 21).

When the Royal (Sapa) Inca moved people from one section of the empire to another, he gave them a new *huaca* of origin (a *huaca* could be a natural shrine from which the ancestors emerged to found their descent groups) (Gose 1992). If the *huaca* was a source of water, a small amount of old water source was ritually poured onto the new *huaca*. If the *huaca* was stone, power was shifted by removing the decoration of the old *huaca* (perhaps a piece of textile) and placing it on the new *huaca* (Zuidema 1982).

In North America, Plains Indian art is recognized as having "a very rigid, fixed art style"; it has been categorized as "an aboriginal art style unaffected . . . by European contact" (Wood 1962, 25). The prehistoric pottery of the five Plains Indian sedentary village people (Pawnee, Arikara, Cheyenne, Mandan-Hidatsa, and Crow) was similar "in the paste, surface finish, technique of manufacture, and form" (27). Also standardized across the five tribes was the shoulder pattern, which consisted of "an embellished band extending from the base of the rim [neck] to the maximum diameter of the vessel [shoulder]" (28).

These tribes did not share a close common ancestor. They spoke distinct languages; the Mandan-Hidatsa and Crow spoke a Sioux language; Cheyenne, an Algonquin language; the Arikara and Pawnee, a Caddoan language. The technique of incising pottery shoulders appears to have been introduced by outsiders. It entered the Central Plains through trade via the medium of Mississippian groups along the Missouri River (Wood 1962, 34).

While diffusion of innovation appears to account for the shoulder incising, the full story is more complex. The technique of incising seems to have been adopted only after metaphorical (or fictive) kin ties had been created. The trading partners who introduced the new technique also created a trade language, often using signs and "a social subsystem, consisting of a network of real and fictive kinship" (Wood 1980, 98). The calumet ceremony, which created the kinship ties, was a "father-son adoption ceremony . . . which enabled members of warring tribes to trade in peace. The Mandan were adopted by a fictitious father, and in turn had adopted sons in tribes with whom they dealt . . . Plains Indian trade was accomplished by barter between fictitious relatives. From a larger perspective, a vast network of ritual relationships extended throughout the entire plains" (Bruner

1961, 201). If the trade partner-kinsman died, the relationship was established with another member of the same group, usually a relative of the deceased. Further, the social relationships of the trading partners were often reinforced by systematic intermarriage between the families of the trading partners (Walker 1967). A similar system of creating metaphorical kinship established around trade and the introduction of new motifs has been described in detail for native Australia (McCarthy 1939).

The point raised by this discussion is that when art styles change in a traditional people, the new style is not adopted cavalierly from anyone. Metaphorical kinship ties had to be established before the styles were adopted, the new ancestors were linked to the old ones, or the old ancestors experienced a rebirth.

## The Trajectory of Visual Art Is Unpredictable

Miller predicts that the evolutionary trajectory of art, like the trajectory of any runaway selection, will be unpredictable (158). Although I agree that the specifics of change cannot be predicted, I would predict that visual art styles will vacillate between art that promotes individualism and creativity and art that promotes kinship and tradition. The conditions that favor traditional art include, among many things, ancestors, strong kinship systems, and traditional moral rules (decorated by art) that encourage honoring the elders and ancestors and being generous to and coming to the aid of all who share ancestry. People who, among other things, have lost or discarded their ancestors, who ignore the elders, and who have fragmented kinship systems will make nontraditional visual art and will value individualism and creativity.

## Ancestors Do Not Guarantee Visual Art

The ancestress hypothesis does not propose that a traditional people will necessarily produce visual art objects, it merely proposes that when they do so the art will identify kinship, literal or metaphorical, and promote kinshiplike cooperation among codescendants. Visual art, because it is attractive, can easily become competitive. It is not uncommon for religions to place limitations on a practice, such as making visual art, that could easily become competitive. Jewish tradition has at times placed a ban on images. Exod. 36:38 describes the construction and decoration of the tabernacle. In Exod. 20:4 and Deut. 5:8, however, the making of any image or likeness of man or beast is prohibited.

## Art Depicts Violence Rather Than Cooperation

A function of traditional visual art is to promote kinshiplike cooperation, but it does not necessarily encourage cooperation with outsiders. By calling attention to individuals sharing common descent, visual art distinguishes those who share a common ancestor from those who do not.

From the ninth century B.C.E., when initial steps toward agriculture were taken, through the reign of the last great Assyrian king, Ashurbanipal, the palaces of Nimrud and Nineveh were decorated with wall bas-relief carvings depicting scenes from Assyrian history. The techniques used to make these bas-relief carvings remained consistent over time; faces were done in profile and were masklike, impassive; and the carvings were stylized (Bersani and Dutoit 1985). Themes of the carvings, however, varied; some depicted such things as King Sennacherib sitting on the throne, irrigation machines and a seeder-plough, a cow with her calf, horses being watered and groomed, women and children in a cart drawn by oxen, a ziggurat, and

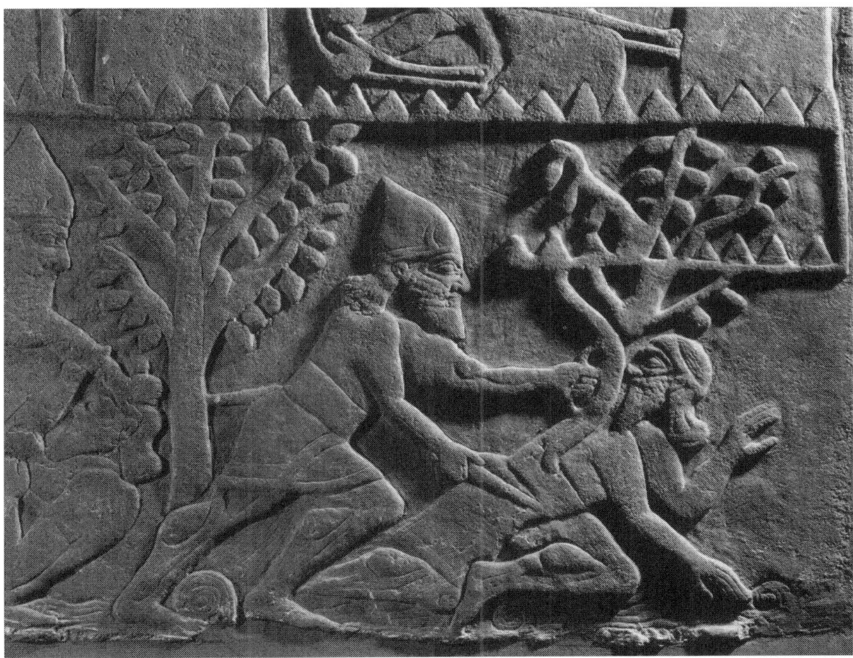

**Fig. 43.** An Assyrian warrior killing an enemy (stone bas-relief), from the palace of Ashurnazirpal II in Nimrud, Mesopotamia (Iraq), circa ninth century B.C.E. Courtesy of British Museum, London, Great Britain and Erich Lessing/Art Resource.

minstrels playing in a garden (Saggs 1984). Other scenes, however, depicted, in gory detail, battle scenes in which the Assyrians flayed, impaled, and burned prisoners (Bersani and Dutoit 1985).

## Visual Art and Associated Rituals

Miller argues that rituals, or cooperative display involving the arts, have high time and energy costs and no possible survival benefit. In his discussion of one initiation ritual, he predicts that its aim is to improve (presumably unmarried) "daughters' mating prospects by demonstrating their wealth, health, family size, and other aspects of familial fitness" (2000, 220). Rituals, myths, legends, totems, and dances, Miller concludes, are used to display the superiority of one group over other groups coming together (220–21). While I appreciate Miller's reference to kinship, as it is one of the few he makes, he is creating a group argument. Members of a tribe or other descent category share rituals, myths, and legends about their common tribal ancestors. They do not make up a group.

Rituals among the Yaqui and the Australian Aborigines involve copying certain movements and decorations used by the ancestors or that depict the ancestors. Traditional visual art, as it involves the stylized copying of behaviors, is itself a ritual. While the production of contemporary forms of visual art may be ritualistic, those rituals are often unique to individuals, perhaps combining old and new rituals, or to particular schools of art. In either case, visual art and associated rituals seem to play an important role in the enculturation of the next generation.

Rituals are associated not only with the identification of codescendants but also with the encouragement of their cooperation. As Howells explained, rituals are "a highly dramatic way of bringing their attention to the qualities of forbearance and mutual help, and of condemning the forces of disunion" (1948, 193). An identifiable effect of religious rituals could be higher levels of patience, forbearance, and mutual help. Keightley describes this link between rituals and kin cooperation in China: "It may also be noted that the invocation of the individual ancestors, fixed within their ritual circle, to mark the passage of time would have regularly served to reassert their existence in the minds of the living. The performance of the rituals, indeed, would have required the Shang to 'make time' and to 'take time' for their ancestors, to pay them attention on a regular schedule that would not be speeded up or condensed" (2000, 51). Needless to say, the rituals would have also required the codescendants to cooperate in the conduct of the rituals. What is interesting about rituals is the cost, the coordination, and the copying. While females may identify males who interest them at rituals, and those males may have bright feathers or be the best dancers, that does not necessarily mean the females will mate with or marry those males. If they do, presumably their communication of ancestry and cooperation with other males, their codescendants, play some role in females' choices.

Clearly, more work must be done to test the ancestress hypothesis, particularly in the area of male decoration. How attractive can a male be without creating a competitive situation with other males? Why are males who head large multinational corporations likely to dress themselves in suits that are, while expensive, likely to be drab or colorless? Why do the women who compete for those males array themselves like brightly colored peacocks? Better tests certainly need to be devised.

# CHAPTER 9

# Modern Darwinian Theory

Although the basis of the ancestress hypothesis is modern Darwinian theory, some of the assumptions underlying the hypothesis, and the evidence presented to support it, do not always seem to fit comfortably. For example, although genes, and often behavior, are assumed to be selfish, there are unselfish mothers. Mothers behave, as Darwin's theory predicts, precisely as we would expect, namely in ways that promote leaving descendants. And, although evolutionary theory predicts that altruism will occur among individuals closely related, codescendants, individuals who share common ancestry (but may not be related at all by current measures of relatedness), are treated preferentially even when those individuals rarely meet. Further, traditions are not aberrant forms of behavior found only when there are powerful males or prelogical or naive people. Traditions are what we should expect to find, as natural selection is, as George Williams noted, more a theory to explain persistence than change.

While evolutionary theory focuses on natural selection working on genes, I focus, as did Darwin, on traits. Although gene-culture evolution is still being debated, traditions (behavioral traits inherited from one's ancestors) are particularly intriguing. While Darwinian theorists count number of offspring, the number of descendants, even distant ones, should be counted, too.

## Darwin and His Theory

Charles Darwin was born on February 12, 1809, in Shrewsbury, England. From 1831 to 1836, Darwin served as a naturalist on a British science expedition around the world. "When on board H.M.S. *Beagle*," Darwin wrote in the first page of the introduction to *On the Origin of Species*, "I was much struck with certain facts in the distribution of inhabitants of South America, and in the geological relations of the present to the past inhabitants of that continent. These facts seemed to me to throw some light on the origin of species—that mystery of mysteries" (1859). The mystery of mysteries that would lead Darwin to his theory of evolution through natural selections was this: Why do organisms—plant and animal species—seem to be so well suited for the environment in which they live? He spent four years and nine months sailing around the world, taking overland expeditions, such as ascending and crossing the Andes, and collecting samples of various species.

Darwin spent only five weeks in the Galapagos Islands. He found them to be ugly and inhospitable, although he recognized that they were a paradise for a naturalist and geologist (White and Gribbin 1995). Of particular interest to Darwin were the finch and the tortoise, as he recognized but did not yet find the fact significant that these species varied from island to island. From Galapagos, the *Beagle* steered a course to Tahiti, then to New Zealand, the Mauritius, Cape Town, back to Brazil, and then home again. He arrived in England in 1836 with 1,383 pages of geology notes, 368 pages of zoology notes, a catalog of 1,529 species preserved in spirits, and 3,907 labeled skins, bones, and miscellaneous specimens (80).

Once back in England, Darwin began to correspond with biologists, geologists, and animal breeders and continued looking for an answer to the mystery. In 1839, Darwin married his cousin Emma Wedgwood, with whom he would have ten children. In 1842, shortly before the birth of their third child, he moved his family into Down House, where he continued to spend hours mulling over ideas about barnacles, coral reefs, orchids, and earthworms as he strolled the shady paths of the estate. He was, as he once described himself, a tenacious man, never giving up until a puzzle was sufficiently resolved. He also was a careful man. While many of Darwin's ideas of evolution were formulated between 1836 and 1842, they were not published until 1859.

The last piece of the puzzle, the answer to the mystery of mysteries, fell into place in 1838, when Darwin read Thomas Malthus and understood the role that competition between individual organisms plays in evolution. Although Darwin was not able to adequately explain why offspring tend to resemble parents (he did not know about genes), he did know that, because children tend to resemble their parents, traits are transmitted from one generation to the next, even though "many slight differences" exist among individuals (Darwin 1859, 59). Darwin argued that the traits present in parents and offspring are a consequence of either natural selection, which produces traits promoting survival, or sexual selection, which produces traits that have survival costs but give individuals an edge in mating.

When Darwin referred to the "preservation of favorable individual differences and variations and the destruction of those which are injurious" as natural selection, he was thinking about survival (91). When he wrote "[c]an we doubt (remembering that many more individuals are born than can possibly survive) that individuals having any advantage, however slight, over others, would have the best chance of surviving and procreating their kind?" he clearly was arguing that survival and reproduction were both necessary (neither was sufficient alone) to account for existing traits (91).

Gaudy and costly traits, such as the exotic plumage of birds of paradise or peacocks, Darwin reasoned, must serve some function. "It cannot be supposed," he wrote, "that male Birds of Paradise or Peacocks . . . should take so much pains in erecting, spreading and vibrating their beautiful plumes before the female for no purpose" (1871). Darwin did not concern himself with explaining why peahens came to like the peacocks' tails; selection would occur if the females liked it and selected the male as a mate because he had the trait. The combination of a propensity for longer and brighter tails that was passed on to offspring, and the peahen's preference for the longer brighter tail (which was also passed on to offspring), would lead, over many generations, to ever brighter and gaudier tails. When Darwin was asked by the Duke of Argyll if he believed that the great beauty of nature was really a chance thing, Darwin responded by saying that as difficult as it was to accept such an idea, he did truly believe that evolution worked without design (White and Gribbin 1995, 275). The beauty of the tail of the bird of paradise was not the consequence of design by a creator; it came from the mundane need to attract a mate. Over time, given both survival and reproduction, "the more favourable traits" came to replace those that were less favorable (91).

## Modern Darwinian Theory

Although Darwin did not know about genes, and therefore focused on traits, we now know that the variation in traits that Darwin noticed is produced by mutations in genes and changes in the environment. Genes and the environment together shape inheritable traits. While mutations are random, natural selection is nonrandom, moving always in the direction of increasing reproductive success.

### *Adaptations*

The traits that Darwin referred to as favorable are now called adaptations, a term used to refer to traits (phenotypic features such as morphological structures, physiological mechanisms, and behaviors) that have an effect or function that lead to the differential reproductive success of the individual organisms that inherited those genes and the traits that, in a particular environment, those genes produce. Adaptations are facultative, meaning that they are conditional and their expression is dependent on the presence of certain environmental cues that were present in the ancestral environment. To understand the function of any adaptation, we must understand its function in the environment in which it evolved. In the case of maternal behavior and visual art, this means looking for its effect or function in the environment of our ancestors. For that reason, this book has focused primarily on people whose life patterns more closely mirror those of more distant ancestors.

Although sociobiologists focus on adaptations, not all observable traits are adaptations. If a trait is due to a recent mutation, natural selection will not have had time to act on it. Traits also may be neutral (having no effect on reproductive success) or a by-product, perhaps carried as a pleiotropic or secondary effect of the development of a trait that is adaptive. Traits may be due to what Stephen J. Gould has referred to as phylogenetic inertia, meaning traits are a historical legacy. For example, the digestive and respiratory tracts cross in the human throat. Choking on food and death from asphyxiation may be a legacy of a distant invertebrate ancestral species and found in all vertebrates since that time (Williams 1992). While a trait's function can change (for example, bird feathers did not initially evolve for flight [Alcock 1998]), environmental change can mean that a trait is no longer adaptive, it may now be maladaptive, perhaps a pathological reaction to the novel environment.

George Williams cautioned us to avoid regarding all the inheritable traits that exist as if they were adaptations: "Adaptation is a special and onerous concept that should be used only when it is really necessary" (1966, 4–5). This does not mean, however, that we should ignore the possibility that a trait is adaptive. It is only by considering such a possibility that adaptations, and their functions, will be discovered.

## Effect as Cause

The task of looking for a trait's adaptive effect is difficult as traits can have more than one observable effect, all but one of which are mere incidental or even perhaps fortuitous byproducts. Williams proposed that "a plausible demonstration of design" (meaning the trait was produced by natural selection) consists of demonstrating that the trait accomplishes its alleged function (its effect) with "sufficient precision, economy, and efficiency, etc." (1966, 254). What one has to ask in regard to any possibly adaptive trait is how it, with precision and efficiency, helped individuals within ancestral populations leave descendants.

Most evolutionary biologists analyze traits based on two complementary levels of cause, proximate and ultimate. Proximate causes of behavior are the immediate causes, such as the hormones and emotions, that are said to influence or cause one to behave in a certain way. If we argue, as many do, that people produce visual art because to do so elicits a pleasurable emotional response, then we are proposing a proximate explanation for why people make art. This does not contradict the view that the behavior has evolved

Ultimate explanations explain why a particular emotional response might occur. An ultimate explanation, thus, would ask *why* visual art came to elicit such a response. It could be argued on a proximate level, for example, that we eat ice cream because it gives us pleasure. The ultimate explanation is that this behavior apparently gives us pleasure because in the ancestral environment sugars and fats were rare concentrated sources of calories that were important for survival and reproduction. Similarly, I am proposing that the proximate effect of visual art, was to use color, form, and pattern to attract attention to objects and messages, thereby making those objects and messages more noticeable and thus more influential. The ultimate effect of art was to attract attention to kinship and ancestry and to messages encouraging cooperation among those so identified. In so doing, visual art played a role in increasing the number of individuals who were identified as kin and who cooperated in protecting, nurturing, and mentoring one's offspring.

The fundamental question this book addresses is why humans made visual art. What was the cause of this behavior? The cause of the behavior of making and viewing visual art is that that effect within ancestral populations helped individuals leave more descendants. While we know that differential reproduction is the effect that promotes the replication of a trait, it is not clear at what point a trait's success should be measured. Williams (1966) wrote that natural selection is more a process to explain persistence than change. This suggests that a trait's success should be measured generations after its origin. If the organism that bears those genes and the trait produced by those genes is successful, that organism becomes a distant ancestor.

## Individual Selection and Selfish Genes

For Darwin (1859), competition was the key to evolution. In every species, numerous individuals are born. As population size tends to remain constant, many more individuals are being born than survive and subsequently reproduce. Darwin assumed that the individuals more successful in obtaining the necessary resources and mates would be more likely to survive and outreproduce competitors. Thus, he wrote that his theory would be falsified if an inheritable trait were discovered that promoted the survival and reproduction of other individuals at the expense of the individual possessing the trait. Such altruistic behavior would promote the survival and reproduction of another individual; future populations would be made up of that selfish individual's descendants, not the altruist's descendants.

The resemblance of parents and offspring, and the variability observed in populations, were explained when Darwin's theory was combined with Gregor Mendel's idea of particulate inheritance. Scholars realized that sexual reproduction leads to genetically unique individuals through the random recombination of large numbers of independently assorting genetic units (Maynard Smith 1978; Williams 1966). As organisms with individualized sets of genes can be expected to have individualized interests, conflict arises. The thinking underlying this was that by shifting our perspective from the level of the individual to the level of the gene, and by thinking of genes as if they had the attributes of individuals (that is, they are capable of doing things in their own self-interest or of being altruistic toward other genes), it is possible to see that natural selection would favor only the selfish genes and select against the altruistic ones. Selfish genes have effects that cause copies of themselves to get into future generations. These effects include influencing their organisms to eat, mate, and raise offspring. Richard Dawkins's selfish-gene theory, however, did not necessarily mean that genes influence an individual to act selfishly (that would have been selfish-organism theory).

Once the focus was placed on selfish genes, evolutionists had to come up with theories to account for the evolution of altruism. As George Williams explained, "How could natural selection, based on the relative rates of reproduction of different individuals, favor genes that cause their bearers to expend resources to benefit their genetic competitors?" (194). The solutions proposed for this riddle were kin selection, or inclusive fitness, and reciprocal altruism.

## Selfish Genes and Unselfish Behavior: Explaining Altruism

### INCLUSIVE FITNESS

When William Hamilton (1964) pointed out that copies of genes are found not only in offspring but also in other relatives, he could explain altruism toward offspring and altruism directed at other kin. A gene for altruism can spread, even if altruists sacrifice their reproduction in order to help other individuals reproduce, if that altruism is directed at kin (who share the same altruistic gene) and if the altruism helps those individuals have more offspring due to the altruism. From the viewpoint of the gene, this seemingly altruistic behavior is therefore selfish. Inclusive fitness refers to the sum total of genetic success through personal reproductive success (or direct fitness) and increased reproductive success of relatives due to assistance given them by kin (indirect fitness).

The number of genes shared by individuals is referred to as the coefficient of relatedness, and the closer the relationship the higher this coefficient. Siblings, on average, share ½ of their genes, while first cousins share ⅛, and second cousins, 1/32. The coefficient of relatedness should be correlated with the degree of altruism; kin altruism between siblings should be more intense than altruism directed at cousins; altruism is not predicted to occur beyond the level of second cousin.

In social insects such as the sterile castes of hymenoptera, sterile worker sisters, due to the haplodiploid nature of sex determination, are very closely related. Sterile sisters, by helping their relatives reproduce, indirectly propagate copies of themselves in those relatives and thus, theoretically, transmit more copies of their genes to the next generation than they would have if they themselves had reproduced.

Although tests of inclusive fitness theory have supported that humans regularly favor close kin, it does not lead to the prediction that kin relationships will be eternally cooperative. Given conflicts of interest, even maternal behaviors, which Dawkins referred to as "the commonest and most conspicuous acts of animal altruism," are seen as contractual, conflictual, and self-maximizing (1976, 6). As Alexander explains, "even parents, among humans, often behave as though they expect their offspring to return certain kinds of favors or assistance to them because they have provided for the offspring" (1979, 54). Offspring, on the other hand, try to manipulate parents into giving more than the parent wants to give (Trivers 1974). As manipulation of parents is often at siblings' expense, relationships between siblings are also conflictual. As Alexander explains, "a gene in an offspring may spread itself even if its effects are contrary to the genetic interests of the parent and the rest of its brood" (1979, 39).

## RECIPROCAL ALTRUISM

Robert Trivers (1972) recognized that cooperation could occur between individuals who are nonkin (individual organisms related beyond the degree of first cousin), and he proposed the theory of reciprocal altruism to account for this cooperation. If a pack of wolves brings down a fawn, there are immediate rewards for all hunters. This kind of cooperation is widespread because of the instant payoff. It is not an example of reciprocal altruism. Reciprocal altruism involves risk, namely a time lapse between giving and receiving. It is a type of social contract in which the donor, voluntarily, is altruistic, as an equal or acceptable return (in prestige, increased resources, or increased number of genes transmitted to the next generation) is expected for the altruistic act.

Trivers recognized that reciprocal altruism involves risk and depends on trust; donors are vulnerable because it may not be in the recipient's self-interest to reciprocate. Given self-interest, one would predict that individuals would try to cheat, collecting the benefits without paying individual costs such as giving alarm calls, sharing food, or even caring for the young (Miller 2000, 314). Consequently, Alexander (1979) argues that reciprocal altruism is more likely to occur if the donated benefits are relatively inexpensive compared to the returns; if altruism is restricted to situations in which a return is immediate, as when a group is faced by an outside threat; if there are policies or written rules enforcing compliance; or if the witnesses who observe the act can exert social pressure on a nonreciprocator. Reciprocal altruism is also more likely to occur if individuals discriminate based on the prior behavior of the recipient. Reciprocity is most likely to occur if there is a long series of interactions between potential cooperators, a situation likely to occur only in small groups in which people have regular interactions (Boyd and Richerson 1990).

## GROUP SELECTION

Another explanation for altruism, which is used infrequently by biologists but quite frequently by social scientists, proposes that the group is a primary influence on social behavior. In 1962, V. C. Wynne-Edwards proposed that groups (a category that could include an entire species) that had population-regulating mechanisms (for example, sacrificing reproduction during tough times to keep the group size low) would survive, while others, lacking those mechanisms, would overexploit the resources in their environment and become extinct. The group, in other words, has a dramatic influence on the behavior of individuals living in the group.

If this group altruism, however, is examined using an individual level focus, the future generations of this particular population regulating group would be made up of the descendants of individuals who continued to reproduce and who managed, by hook or crook, to get resources to do so. Williams (1966) pointed out that the self-interested individuals would accumulate resources, thus promoting their own survival, while altruists would promote the survival of the recipients of their generosity. Group altruism, consequently, is unlikely to be produced by natural selection.

While complex models of how group selection might occur have been postulated (Wilson and Sober 1994), most evolutionary biologists today accept individual selection as the primary force for evolutionary change. Group selection, although theoretically possible, is unlikely to occur, as the conditions that would make it possible are so stringent. Groups would not only have to retain their integrity for long periods but also differ in genetic constitutions in ways that affect fitness (Alcock 1998). Despite the weakness of group selection theory, many hypotheses related to human cultural behavior, including traditions of visual art, religion, and kinship amity, turn to group selection arguments, because there seems to be no other explanation that can adequately account for the data. The ancestress hypothesis attempts to explain cooperative behaviors without making reference to a group. Individuals in each generation make the choice to give importance to or ignore their ancestors and traditions. The point is not that ancestors live forever among us but that when they and their traditions are important or are dismissed certain predictions can be made about the social behavior, including the visual art, of the individuals who accept or reject them.

# CHAPTER 10

# Reconciliation

## *The Problem of Definitions*

Natural selection is a process that logically occurs if there is variation among individuals in a population, if that variation is inheritable and/or replicable, and if individuals, due to inherited variation, experience differential reproductive success (see Alcock 1979). If Darwinian theory is used to explain the arts, unselfish mothers, or moral systems that promote cooperation among codescendants, it must be done so by reference to individual strategies and competition. I use the word *reconciliation,* which comes from the Latin word *reconciliare,* meaning to make consistent, not as a criticism of Darwinian theory. My concern is related to the way we currently define certain terms and the assumptions that the definitions have generated. Definitions must be emphasized to avoid creating assumptions or building hypothesis that, while attractive, are fundamentally nontestable.

*Reproduction* is one word that seems to have various definitions. Does reproduction refer to the act of copulation, success in which may not lead to the meeting of an egg and sperm? Does reproduction refer to conception, childbirth, lactation, childcare? Is one reproductively successful when one's child becomes a parent? Is one successful if that child's child becomes a parent? We need to distinguish between strategies that may work in the short term (immediate reproduction—copulation, conception), although they may not be successful in the long term, from strategies that may limit immediate reproduction in the short term (in that they limit, as one example, the number of mates we can attract or children we can produce) but may be more likely to make us ancestors.

Further, we refer to cooperating groups as *coalitions.* In common usage, *coalition* refers to temporary social relationships. What seems to distinguish enduring relationships from temporary ones is not only the larger investment that they require (involving such things as patience, generosity, watchfulness—social skills often associated with good parenting—which are at the cost of our survival and immediate reproduction), but also the fact that an enduring relationship links us, at least theoretically, to the enduring relationships of the other individual. Put simplistically, one inherits, as one example, one's parents' enduring relationships, and even after the parent's death, those individuals are likely to maintain an interest in the offspring's well-being. This means that the number of individuals available to provide assistance can be large; however, altruism, as this example shows, becomes more complex as it involves nonkin, yet is morally imperative, and as any return on an act of altruism is likely to be returned to another individual, not the original altruist.

A second term, the accepted definition of which may have wider implications, is *competition.* Competition, for Darwin, could refer to the behavior of the roots of two plants, one of which reaches deeper into the soil and obtains more water than does the other. This organism, due to its longer roots, is more successful in obtaining the resources necessary for its survival and reproduction. Thus, we can say that this organism has outcompeted the other organism. Competition does not in any inevitable way imply aggression or violence, although we often use the word as if it did.

Underlying our ideas of both competition and enduring cooperation is the definition of the word *social.* Thus, this discussion starts by examining the definition of social behavior. One problem that may have inhibited our ability to understand social behavior is our assumption that it refers solely to an interaction between individuals.

## What Is Social Behavior?

Darwin, according to most descriptions, was a social man. He loved his wife, found joy in his children, was involved in their lives and education, and appreciated the extended family that was important to his wife. Although his mother died when he was young, he honored his father and was influenced by his grandfather. He liked to correspond with his friends, and he liked having them come to carefully scheduled visits. The word *social* in reference to Darwin means that he behaved in certain ways.

When George Williams initially wrote about social adaptations, he focused on "the less common interactions that do seem to be cooperative and benign" (1966, 193). The word *social,* as Williams used it, referred not to just any interaction but to a particular kind of benign or cooperative interaction. A review of current literature, however, indicates that ethologists regularly use the word *social* to refer to *any* interactions between two organisms. Social behavior, thus, can involve aggressive interactions, such as territorial defense, as well as more amicable patterns of courtship and parental care. Based on this assumption, ethologists seem to contrast the word *social* with the word *nonsocial* or *asocial,* using these terms to refer to individuals who do not interact, are solitary even when they live, as do ground squirrels (*Spermophilus mexicanus*), in close proximity within a colony (Vaughn 1986, 390). The cheetahs of the Serengeti plains are referred to as asocial (Caro 1994). Because nonsocial is a category that includes a number of mammals, including some marsupials, rodents, insectivores, and many carnivores, any interactions between a mother and her offspring, short term or enduring, are presumably being ignored.

We might use the word *nonsocial* to refer to human behavior (hermits, for example, might be referred to as nonsocial); however, when speaking of human behavior we generally contrast the word *social* with the word *antisocial,* meaning that the individual rejects or lacks the capacity for social interaction, thus behaving in a self-interested fashion that is likely to be at the expense of others. Since we contrast the term *social* with the term *antisocial,* then the word *social* in everyday usage has an opposite and we have something with which to contrast the word in order to test the definition's accuracy. If the word *antisocial* is used to refer to the behavior of individuals who do things that are self-interested and at the expense of the interests of others, what then is social behavior?

Social behavior does not require an interaction, and, further, humans can have interactions that are not referred to as social (Steadman 1994). On the Cherokee Trail of Tears, the wife of Chief John Ross "died a martyr to childhood, giving her only blanket for the protection of a sick child" who was not her close kin and may have been a stranger (Burnett, 1839). If a stranger covers an unconscious child with a blanket, or a mother soothes the brow of sleeping child, or someone sends an anonymous donation to charity, these are referred to as social acts. Certainly they are not referred to as asocial or antisocial ones. Because there is no response (the second individual is unaware of the actions), these behaviors, while social, do not involve an interaction. Humans can be social without interacting with another person.

On the other hand, we can have many interactions that normally would not be referred to as social. We can have, for example, an interaction with bacteria or a virus, or with a mosquito, or with a dog that bites us, or with a thief trying to rob us, or an enemy attempting to kill us. We usually would not use the word *social* to refer to any of these interactions. If we accept this argument, several consequences follow. First, if social does not necessarily involve an interaction and interactions are not necessarily social, interaction is not the proper focus of any study of social behavior. Second, although Miller and others attempt to explain social behavior by $r$, mate seeking, sexual behavior is not necessarily social. Rape is a form of mate seeking; all would agree it is an interaction; few would call rape a social activity. In his book on World War II, novelist Moritz Thomsen makes it clear that sexual encounters are not necessarily social: "Standing up and screwing in a telephone booth was so quick, exotic, and degenerate, so absolutely devoid of any human tenderness, so essentially inhuman, that I could never connect these infrequent and fugitive five-minute periods with myself or my emotions. They had nothing to do with me; they relieved an awful tension; they left me guiltless; they cost me three pounds" (1996, 390).

Another consequence is that social behavior seems to refer to behavior that has certain unique characteristics. If we put together the stranger, friend, mother, and anonymous donor, what social seems to involve is one individual's vulnerability and altruism or generosity. Social behavior thus involves risk, in the sense that when we are social we let our attention wander away from protecting our own interests while we attend to another's interests, using resources we could have used to promote our own survival and reproduction to promote another's survival and reproduction.

What these examples describe is good parenting behavior. Most of us accept that parental care is social, that it is a K-strategy behavior, and that it is always at the expense of having more offspring. Parental behavior has high costs and is asymmetrical, in that resources tend to flow from parent to child and there is no necessary response or return on that investment. Trivers has argued that from an offspring's point of view, resources invested by parents in one offspring, at that offspring's encouragement and manipulation, are at the expense of that offspring's potential or existing siblings. This behavior is self-interested from the viewpoint of the offspring, as more of the indulged sibling's genes go on into the future at the expense of its siblings. From the mother's point of view, however, something different is going on that may be crucial to our understanding of social behavior. An increased investment in one offspring, from the mother's point of view, is at the expense of *her* other actual or potential offspring. More of *her* genes would be transmitted to the next generation if she invested time and resources in producing more offspring herself rather than initiating reproduction late in her life (because of adolescent subfecundity), terminating reproduction early (because of menopause), and using large amounts of resources to help children and grandchildren survive and reproduce. When a mother helps her child reproduce by investing resources in that child, 25 percent of the mother's genes go forward, instead of 50 percent, which would have gone forward if the mother had reproduced (Steadman 1995).

While it may (or may not) be true that by investing more in each offspring a female increased the likelihood that offspring would survive and reproduce, this does not imply that she maximized the number of offspring that she could produce. David Lack discovered that female birds adjusted the number of offspring they produced by accurately assessing the environment and their own health and capabilities from year to year as conditions varied. Traditions, however, do not change from year to year; generation after generation they shape the form that parental care takes.

If a mother has more than two offspring (which is usually the case in humans), from the mother's point of view it is adaptive for her offspring, siblings to one another, to cooperate. Without cooperation, the transmission of traditions not only would be impossible, but interactions would be potentially dangerous; given the large maternal investment, it would be costly for a mother if one of her children were to kill another. If a mother has more than one child, there should be a corresponding selection for sibling behavior (Steadman 1995). Sibling behavior, following from the above, would involve a selection for restraint. Among humans, the encouragement of this restraint is parental. To the extent that grandparents are involved, this encouragement is grandparental; it is traditional. What is interesting about restraint is that it is a form of social behavior that is at the expense of the restrained individual's survival and reproduction.

Most of us accept that K behavior involves not only vertical forms of cooperation (parent to offspring), but also horizontal (sibling to sibling, mother to father) cooperation. Cooperation, following from the definition of social, is a form of social behavior that involves mutual vulnerability and generosity. To provide an example, when Captain James Cook first arrived in Hawaii, he was treated well (as a deity, the records say). When he returned to Hawaii, he was killed and dismembered. While we might refer to the first meeting as a social event, Captain Cook (and even the Hawaiians who had to fight against men with more sophisticated weapons) may not have used these terms to refer to the second meeting. When the Hawaiians first welcomed Captain Cook, both Captain Cook and the Hawaiians took a risk and both behaved generously. The Hawaiians, by welcoming him, set into play a series of events that, as history would show, largely has led to the loss of their ancestors, the breakdown of their kinship and descent systems, and the loss of much of their culture. Captain Cook lost his life.

If this argument on K-strategies versus r-strategies is correct, social behavior in mammals seems to have evolved out of motherly care. Motherly care, we accept, is a K-strategy behavior. K behavior at the expense of survival, because it involves making an investment in another that limits one's own survival. Further, because an r-strategy can always outreproduce a K-strategy, K-strategy behavior also cannot be explained as promoting reproduction, as it is at the cost of mate seeking and mating. K is always at the expense of r-strategy behavior, just as r-strategy behavior is always at the expense of K. What this means is that social behavior cannot be explained as promoting either survival or mating, because it is at their cost. Yet K has positively been selected for; mammals would not be around if it had not. Being social, as Williams recognized, involves sacrifice. Humans created a strategy that apparently did not work in the short term, as it involved sacrifice and led to fewer offspring, but it must have worked in the long term, over time (as it led to more descendants), regardless of conditions that varied from year to year.

## The Problems of Selfishness

The word *selfish* is an odd word to select to refer to genes. Selfish, according to the *Random House Unabridged Dictionary of the English Language,* means "devoted to or caring only for oneself; concerned primarily with one's own interests, benefits, welfare, etc., regardless of others." *Selfish,* in other words, is a term commonly used to describe behavior; basically, it is a criticism of certain behavior, namely behavior that promotes one's own interests and is at the expense of another's. Other people use words such as *good* and *bad, moral* and *immoral, selfish* and *unselfish* for one purpose: to encourage or discourage these behaviors. Others, most often our ancestors, have determined what is good or bad; fundamentally, words such as *good* and *bad, moral* or *immoral* come from the past. Good and bad are not objective definitions; they do not refer to an objective reality. The point of those words is that ancestors used them to encourage altruistic behavior and discourage selfish behavior within a set of their cofollowers, their codescendants. Selfish behavior hurts codescendants.

The interest of a selfish gene, for Dawkins, was to promote its own persistence in future generations. While it is true that certain genes persist because of the particular effect they have within a particular environment (an environment that is complex, including even the other genes within that organism), all that genes really do, as Donald Symons (1979) pointed out, is influence cells. Genes are not really designed for future effects. Genes are found in individuals in any given generation because the ancestors of the living individuals left descendants. If the environment changes, the gene may no longer have this same effect.

Natural selection, however, acts on phenotypes, not genotypes. This means, in terms of traits (for example, anatomical, physiological, behavioral), that the effect of the trait promotes the trait's persistence. Effect, in sum, is cause. Genes for color vision, for example, persist because of their effect, such as being able to select ripe fruit; it is adaptive and promotes the survival and reproduction of the individual who inherited the trait. Genes for social behavior persisted and became widespread because the effect, whatever that was, apparently outweighed the many costs involved in social behavior.

Selfish-gene organisms are not necessarily selfish organisms; this is not the selfish-phenotype theory. Social humans do not necessarily behave selfishly. Humans, however, often are described as if they were selfish. Dawkins warns that if we wish to build, as he does, "a society in which individuals cooperate generously and unselfishly towards a common good you can expect little help from biological nature. Let us try and teach generosity and altruism, because we are born to be selfish" (1976, 3). This inconsistent use of the word *selfish* is problematic.

## Good Genes, Traditions, and Good Mothers

When we refer to someone as a good person, we generally mean that that person behaves in ways that seem to be altruistic or unselfish. Darwin, however, used the word *good* to refer to traits that were produced by natural selection and that had a positive effect on the organism's ability to survive and reproduce. Darwin's assumption was that good traits persist because they promote the survival and reproduction of individuals who inherit the traits (1859, chap. 4). Darwin's use of the term *good* meant, as Williams (1966) noted, that the trait worked over time. Darwinists today use the word similarly, to refer to such things as good genes. Females looking for a male with good genes are looking for a male who is healthy, parasite free, able to produce viable offspring, and (when confusion is possibly due to the presence of closely related species) of the appropriate species or subspecies. By mating with such a male, copies of a female's genes are more likely to be transmitted into the next generation. The term *good,* as used by Darwinists, has nothing *necessarily* to do with unselfish or altruistic behavior.

A mother's interests, evolutionarily speaking, are to get copies of her genes into the future. While the mother's genes may be aimed at getting themselves into the next generation, the mother cannot help the genes do so by being devoted to or caring only for herself. A huge part of a mother's interest is the well-being of her offspring. This may be especially true if the mother developed in an environment that included traditions.

Traditions are culturally inherited traits that persist, transmitted from parent to child. One tradition that is widespread and apparently ancient encourages women to be good mothers; that is, to put their children's interests first by giving up their own personal comfort to ensure their children are comfortable, giving up their own food to ensure their children are fed, and giving up their own carnal (meaning desiring meat) or hedonistic desires (for example, the desire to have time for oneself, a full stomach, a room to oneself, etc.) to be patient, generous, and dutiful.

Edel and Edel argue in their cross-culture study of ethics that "Mother, take care of your child" is a "universal imperative" (1959, 34). Given the needs of children, "the need for maternal care is an absolute: children need the sheltering care of a mother if they are to grow up at all . . . no society that has failed to provide all these elements could possibly survive; and on the whole it seems this must be provided by a mother or a very nearly equal mother surrogate." Edel and Edel tested their argument by looking at the Marquesas and Mundugumor, where women reputedly refused to bear children to avoid spoiling their figures, and mothers practiced infanticide "to spite their husband's families," were harsh in disciplining their children, and rejected and handled their infants roughly (114). To discover whether or not these mothering behaviors were new behaviors or traditions that the elders truly tolerated, they asked: "What do the old people think of the rejecting and neglectful young mother?" "Is she punished for her behavior?" Poor mothering behaviors, they concluded, were not tolerated. Further, the mothers' "cruelty and destructiveness were so great that being a tiny group they must surely have killed each other off if they had continued in the same way much longer. . . . We must ask of any set of data which appears to contravene or limit the universal [be a good mother], how viable is this society in fact?" (41).

Mothering behaviors, they claimed, are so important that they can be seen to have an "absolute structuring effect" upon morality and serve as the foundation for restrictions and positive ideals and virtues (114). Being a good mother is held to be so important that mothers, cross-culturally, generate moral sentiments. What Edel and Edel are arguing is that while some maternal strategies are good in the sense that they work through time, others, such as those found at one time in the Marquesas, are not.

Many readers are already familiar with the importance of the naturalistic fallacy: the mistake of deducing what is good in a moral sense from what is natural or what has been favored by natural selection. That is, if someone argues that rape may be good in the evolutionary sense of having been favored by natural selection (see discussion in Thornhill and Palmer, 2000), this in no way implies it is good in the moral sense. This is indeed a fallacy, and such assumptions must be avoided.

Avoiding the naturalistic fallacy also requires remembering a third sense of the word *good*. When I refer to good mothers or good mothering behavior, I am using these words in accordance with how they are generally used in our society to describe certain mothers. For example, people would tend to say that a good mother tries to do what benefits her children; the bad mother indulges her own desires at the expense of her children's well-being. This use of the term *good* happens to often coincide with what is considered morally good behavior. As readers with knowledge of evolution will realize, this use of the term also happens to often coincide with the behavior that is good in the evolutionary sense. That is, good mothering behavior in the everyday sense tends to increase the survival and reproduction of offspring, which is good in the evolutionary sense of being favored by natural selection (although this may not always be the case).

Mothers described in traditional societies as being morally good may merely be those who copied the strategies of their mothers, and those strategies happened to be crucial in promoting their ancestors' long-term success in leaving descendants. By referring to such behaviors as morally good, ancestors would have encouraged their daughters to behave that way, thereby promoting the replication of the traditions, their daughters' reproductive success, and their own and their ancestors' long-term success. *Dynastic strategy* refers to behaviors considered good morally and good evolutionarily speaking.

Success in the past, however, does not predict future success. Therefore, none of these uses of the word *good* constitute my own, or an inevitable, actual moral judgment of what is good. Readers are free to decide for themselves whether or not they consider good mothering behavior to be good or bad, just as they are free to decide whether behavior that has been favored by natural selection is good or bad. Indeed, they are free to conclude that good mothering behavior that has also been favored by natural selection is bad, and that the bad mothering behaviors that were selected against are the very best behaviors in which a human could engage. Indeed, any possible combination of the three senses of good and bad is possible.

That dedicated mammalian mothers tend to become ancestresses may seem obvious, but it is emphasized for several reasons. First, as this role is widespread, it presumably is ancient. Second, as much of mothering is learned (how else do we explain the popularity of how-to books on child rearing?), and acquired by copying, females apparently need to be taught to be a good (or bad) mother. Third, the fact that there is an explicit (or even assumed) rule suggests that human females not only need to be taught to be good mothers; they need to be reminded.

## Defining Culture and Traditions

While anthropologists have had more than a century to reach a consensus on how the term *culture* should be defined, they have yet to do so. Perhaps for that reason, many anthropologists today do not bother trying to define the term, while others continue to accept the definition coined by Sir Edward Burnett Tylor at the very beginning of anthropology: "That complex whole which includes knowledge, belief, art, morals, law, custom, and any other capabilities and habits acquired by man as a member of a society" (1960, 1). As Tylor's definition is actually a mere laundry list of examples, it is not a useful definition. What then is culture?

Anthropologists tend to agree that culture is something that is shared. This means that idiosyncratic behavior is not cultural (see Ember and Ember 1996, 195). Further, they agree that culture is learned, or acquired from others, and can be transmitted horizontally or vertically (Cavalli-Sflorza and

Feldman 1981). Innate behaviors are not cultural behaviors (Ember and Ember 1996, 196). We typically use the word *tradition* to refer to a category of cultural behaviors that are ancestral, transmitted from parent to child across generations.

While some anthropologists focus on culture as behavior, many focus on culture as ideas or attitudes (see Keesing and Strathern 1998), and evolutionary biologists often seem to follow this position. Memes seem to exist in the brain, as do symbols, attitudes, ideas, and rules, and they perhaps have an effect on survival and reproduction (Knight, Dunbar, and Power 1999, 9–10). While ideas (and memes) may be important to humans, I follow others (Palmer and Steadman 1997) in assuming that what seems essential to culture is not just that it is learned and shared but that it is acquired from another individual and potentially transmittable to a third. What is inside our skulls is inaccessible to others. Only if a cultural trait is exhibited can it be copied. What can be copied is what people do and what they say. Talk about ideas or beliefs, but not the ideas or beliefs themselves, can be copied and transmitted to the next generation. Our readiness to copy others, to even copy what they say when there is no empirical evidence supporting that what they say is correct, is based presumably on genes and therefore is subject to natural selection.

## Traditions and Natural Selection

Traditions, anthropologists tend to agree, are behaviors coming from the past. As humans lived in small groups of kin, traditions were behaviors transmitted from parent to child, often over many generations. While I find the persistence of traditions intriguing, many, including Sarah Blaffer Hrdy (1999) find them to be theoretically troublesome. Although each of our offspring inherits about 50 percent of our genes, all of our traditions are transmittable to all of our offspring. Yet traditions lead parents to bind their daughter's feet, sequester them, and sew their genitals closed. Traditions led parents to insist that their daughters adhere to rules of modesty and obey taboos isolating them when they were menstruating. Traditions led parents in many parts of the world to circumcise their male infants and Australian Aboriginal parents to push their sons into a ten-year initiation involving dental ablation, subincision, scarification, food taboos, and a pilgrimage into enemy territory. Traditions led elderly Eskimo to die alone on ice, and they encouraged soldiers to die fighting for their people and ancestral lands.

Traditions are theoretically interesting precisely because they can involve sacrifices of freedom, time, resources, and comfort, yet they can persist for a very long time, hundreds or even thousands of years. Although we are a species that seems to be unusually good at serving our own needs and desires, traditions encourage restraint of behavior. When we follow traditions we are not free to do as we please. Further, we expect parents to be concerned about their offspring's well-being; yet parents are the ones encouraging the often-painful traditions. We expect to find conflict between parents and their offspring (Trivers 1974; Alexander 1974). Traditions, however, imply profound parental influence over offspring.

The ability to understand traditions is inhibited by the imprecise way the term is used. It is used to refer to anything that has occurred in the past, or that has lasted more than a few years, or that has been borrowed from people who live a non-Western lifestyle. A more precise definition specifies that traditions are behaviors transmitted from parent to child across generations. Thornhill and Palmer explain it parsimoniously: Traditions are phenotypes (2000, 25). They are enduring cultural behaviors that imply parent-offspring similarity. Traditions, according to Thornhill and Palmer (2000), occur only when genes interact with a multitude of things from the environment, including ancestors who perform the behavior so it can be copied by the next generation. While it is popular to think that our behavior is triggered in response to certain environmental cues, and, for example, that mothers are more likely to neglect certain infants or all infants in certain environments, traditions are strategies that endure.

When traditions are seen as traits inherited from ancestors, it is possible to appreciate that human behavior has not been the product of an undefined and unspecified group, even though this is a common claim of anthropologists and implied in many discussions of culture. Our ancestors, not our group, have been the primary influences on the way we behave. For much of human history, our behavior has been the product of our ancestors (Steadman 1995). The transmission of behavior from ancestor to descendant differs significantly from a horizontal transmission between peers. Behaviors transmitted to children can have multigenerational effects, and for this reason parents are usually careful about what they say and do around children.

Given that traditions are inherited (children copy them from their parents) and can persist for generations, it is at least theoretically possible that behaviors, such as those found among the Shakers, that prevent or discourage copulation and reproduction would eventually die out along with the individuals who practiced the traditions. The Shakers left no biological descendants; they transmitted no behaviors to their heirs. The persistence of what we might call a Shaker community (meaning in this case individuals who shared a religion as well as many fruitful practices) was not due to the dynastic success of the original founders. The fact that we could find, generations later, individuals who identified themselves as Shakers can be explained by their practice of adopting orphans and raising them as Shakers.

It is also theoretically possible that traditions that promote the survival and reproduction of one's descendants would persist because of their effects. Miller (2000) toys with this idea when he writes that during the Pleistocene, "one generation's experiences of courtship and parenting would have been much more relevant to the next generation" (221). What Miller does not appreciate is that traditions can last much longer than a generation or two and that the *massive* accumulation of traditional behavior is as unique to our species as is the large brain.

The key to traditions as adaptations is this: As all of our offspring can inherit our traditions, traditions should respond rapidly to natural selection (Palmer and Steadman 1997). Further, traditions can be expected to show evidence of adaptation *only* to the extent that both the genetic and the environmental influences on that behavior have been replicated across generations for the long periods of time needed for effective selection (Thornhill and Palmer 2000). Success in an evolutionary sense means that one has become a distant ancestor; the success of inherited traits can only be measured in distant generations of kin (see Dawkins 1982, 184; Palmer and Steadman 1997). Traditions, by definition, are stable strategies that, as Hrdy noted in regard to the tradition of monogamous marriage (which limits individuals to one partner), persist because they raise "survival prospects for offspring" (2000, 258). Such traditions promote not just an offspring's survival and reproductive prospects, and his or her offsprings' survival and reproduction across generations, but, in so doing, the traditions promote the ancestors' success in leaving not just offspring but also descendants.

Several issues seem to have influenced the failure to recognize the importance of ancestors and traditions. First, traditions are of minor importance in the United States. Second, our theoretical focus on change in response to environmental contingencies has made it easy to ignore the remarkable persistence that was occurring in human cultural behavior. Third, anthropologists often dismiss ancestors as mythological because the precise steps connecting the living with that ancestor cannot be identified. While a number of people have a prohibition against speaking the name of the dead, it is not always true that precise links are not identified. Some traditional people recite or chant the long list of ancestors that connect them back to the original ancestor. It is not clear why the regularly heard claim that the ancestors can influence the living was ignored or why it did not seem important that the ancestors were credited with creating culture and giving their descendants a blueprint for how to live life.

Perhaps ancestors and traditions were ignored because we often assume that culture is something humans use to acquire personal power and influence. Ancestors, thus, would be seen as important only when the lineage head began to possess increasing authority (Keightley 2000, 99). Ancestors, however, are important and ancestral rituals are complex in acephalous societies. Even in hieararchical societies, such as that of the Inca or as found in Samoa, ancestors can influence the behavior of their descendants, including the behavior of their direct descendant, who is referred to as the living ancestor or the living god. It is when traditions that encourage restraint are lost that leaders are free to define their position.

## Cooperation in Tribes, Clans, Etc.

In Australia, kinship plays a central role in social life. It is, Berndt and Berndt write, "the articulating force for all social interaction" (1977, 90). In these social interactions, the same kin terms that are used for close kin are used for distant kin or metaphorical kin. Robert Tonkinson writes that there is "no terminological distinction made between close blood relatives and distant kin, so, for example, a person generally behaves toward all those women termed 'mother' in a similar patterned way" (1978, 44). Even though a male will be well aware which mother is his biological mother (genetrix) and will favor her, he has to cooperate with his other mothers, some of whom he rarely sees and others who are strangers to him.

Every kin term "connotes a complex of ideal behaviors" (44). Children learn about the kinship system, he claims, from their grandmothers in particular, who tell them stories about the proper behaviors and about the obligations associated with kinship (45). Children also learn by observation that their grandmothers, for example, are treated differently than an aunt and that uncles are treated differently than a cousin. By the time children are adolescents, they apparently follow the prescribed kinship behavior without thinking. Tonkinson writes that he has "never heard Mardujara express resentment or frustrations at the restrictions that their kinship system places on them. Instead, people talk with satisfaction about the good feelings that come from being surrounded by so many others who are 'one family' and 'one people' with them" (45).

Santos Granero (1991, 246) explains that the Amuesha of Peru's tropical forest refer to all Amuesha as a "big family" and explain that "to ignore the obligations that one has for one's Amuesha fellow is, in the last analysis, equivalent to ignoring the duties intrinsic to kinship or affinal ties." These obligations, he writes, are for "unrestricted generosity" to large numbers of individuals.

The Bedouin of Cyrenaica claimed that their tribe was one large family (Evans-Pritchard 1940). By classifying all members of their tribe or clan as relatives, they encourage certain cooperative behavior (Barnard 1999, 57). The Lugbara, Middleton (1960) writes, had about sixty clans names, each of which identified on average about four thousand patrilineally related kinsmen. As the Lugbara were exogamous, members of a clan were prohibited from marrying one another. Each individual, consequently, had two sets of patrilineally identified relatives (through his father's father and through his mother's father), meaning that he had about eight thousand identifiable kin. Further, as all of the females in his father's clan could be identified (by virtue of their mother's patrilineal identification), the average individual was able to identify by clan name some twelve thousand relatives (Steadman 1994).

Navajos, who introduce themselves by providing the names of their maternal, paternal, and grandparental clans, can identify, through those clan or descent names, hundreds of individuals as clan members. They use kin terms, such as mother, father, brother, and sister, to refer to these clan members and kinshiplike altruism is to be extended to these kin. According to Trudy Griffin-Pierce, "Kinship is so important in Navajo culture that the worst thing that can be said of someone is, 'He acts as if he didn't have any relatives.' Conversely, it is common to hear *naat'áanii* [headmen], admonish others, 'Act as if everybody were related to you'" (Kluckhohn and Leighton 1962, 100). In Scotland, clan names and dress were inherited and identified descent. Traditions of clan hospitality specified that all members of one's clan, even strangers, were to be treated as if they were close kin, and hospitality was even to be extended to members of other related clans.

This pattern of altruism, observed so often by anthropologists, has been given a name. The best-known name is the "axiom of kinship amity," coined by Fortes (1969, 232). The cost of kinship amity, however, is vulnerability, as we know from the infamous massacre of the Macdonalds by the Campbells at Glencoe. Further, participants in the metaphorical kinship groups of the Indians of the Plains may have exposed themselves or their descendants to smallpox, if this disease was indeed carried by kinsmen traders.

Despite claims that kin selection can explain cooperation among individuals who are members of clans and moieties, it cannot do so (see Maschner and Patton 1996, 94–95, for such a claim). Individuals in so-called descent groups are not closely related, as is required by kin selection theory. While kin selection can explain the cooperation observed in other species that do cooperate in groups larger than the family (bees, ants, termites, the naked mole rat) (Wilson 1971; Jarvis 1981), the individuals in these species are closely related. Human clans, moieties, phratries, and tribes (the so-called descent groups) are comprised of hundreds or even thousands of individuals who claim common descent but who, depending on the distance or the number of generations leading back to the common ancestor, are not closely related at all.

Reciprocal altruism cannot explain this cooperation either. First, there is not necessarily a series of interactions. Generosity is to be extended to strangers who are identifiably one's codescendants. A return on any investment might not occur for generations. Reciprocation, consequently, would benefit a distant descendant of the donor. The mandate of favoring kin, Palmer and Steadman (1997) argue, is prescribed by tradition. The system depended fundamentally on traditions that specified duties, as well as on trust that all would live up to those duties. Most clans and tribes had no policies or written rules enforcing compliance. Often, as many of these groups were acephalous, there was no strong leadership with the power to enforce compliance. There are few ways, except for informal ones, such as gossip and ostracism of encouraging, for forcing compliance. The punishment regarded as the most severe was banishment, the loss of all social ties.

Group selection is not likely to provide an explanation for kin generosity either, even though natural selection theoretically could work at the level of the group. The conditions necessary for group selection to occur make it unlikely (Lewontin 1970). Further, and more importantly, clans, just as other descent categories, are not residential groups. Clans are exogamous, meaning that members move out. A daughter, for example, might move to the residence of her husband's family; a son, to that of his wife. Further, clan members live scattered among various residential groupings. Clans thus do not form the type of group necessary for group selection to occur. The same can be said of other descent categories such as tribes (Palmer et al. 1997).

While I have made kinship and descent systems sound fairly simple, they are more complex than most scholars appreciate and certainly more complex than the simple descriptions provided here. To give an example, kinship and descent systems among the Australian Aborigines are so complex that anthropologists have yet to completely figure them out. We do know that Australian clans are associated with totems, or clan symbols, which often are animals, but which can be either animate or inanimate things (Elkin 1933, 1964). This totem, which is used frequently in their visual art, often is said to be the clan's ancestor. Members of these totemic groups, Piddington (1950) explained, could not change their membership (you were born and died a member of the same clan) and they shared obligatory rules of behavior, which included rules regarding how one should act towards one's totemic kin and how to treat totemic objects, including a prohibition on eating the totemic species. Rules also stipulated the use of special terms of address (for example, using kinship terms to refer to one another) and the use of similar decoration or badges, such as the totemic badge.

Elkin (1933) claimed that there actually were heterogeneous forms of totemism in Australia. Cult totemism, Elkin explained, was determined by the child's place of birth, while the dream totem was revealed to the future mother when she felt the first signs of pregnancy. In other words, he implies that these are not inherited; they are not descent categories. A deeper look at the situation, however, reveals a different story. First, the dream totem seems to provide the child with a name and a protective spirit, as is seen in other places. Personal names are identifiers; we find them all over the world. Second, although Elkin claims that the cult totem is determined not by descent but by the place of birth, births can be arranged to occur in a specific place associated with certain ancestral totems. There is, Tonkinson writes, "a strong preference for children to be born somewhere in or near the estate of their father so that they both will share the same ancestral totem" (1978, 51). As fathers arrange the birthplace, this is what generally occurs. If a child, due to adverse circumstance, is born away from the appropriate place, the parents can claim that even though the child was not born on the estate of their fathers, he or she was conceived in that area. The child is thus a member of the same clan as its father.

This cult totem (which Tonkinson refers to as the ancestral totem) is inherited patrilineally. It is the only totem that a male can inherit from his father and transmit to his son. Patrilineal inheritance, however, clearly did not mean that a male was cut off from his mother and her kin. Elkin writes that the sister's son respected the cult totem of the mother's brother, or, in other words, the nephew respected the maternal uncle's cult totem. The uncle, or mother's brother, and his sister (the nephew's mother), however, would both inherit the same cult totem. The son apparently inherits some rights to this clan totem through his mother because he can use the emblem and attend the rituals of the clan that his mother and her brother share. In short, this male is not inheriting his uncle's cult totem, he inherits the totems (and descent group affiliation) of both his mother and father and is associated with both patrilineal and matrilineal kin through the totemic rituals.

The answer to the puzzle of how groups might influence human behavior is simple and is related directly to the ancestress hypothesis. While we clearly are influenced by those around us, throughout most of human history and prehistory, ancestors, not groups, have been the primary influences on behavior (Steadman, Palmer, and Tilley 1996). Individuals who shared common descent shared traditions that they inherited, through their own parents, from a common ancestor. These descendants may live together or they may be widely dispersed; they may all be members of a matrilineage or a patrilineage, or they may claim affiliation with a patrilineage only through their mothers or grandmothers. An individual can be simultaneously a member of more that one group (a matriclan and a patriclan, for example). Members of a clan or moiety may interact regularly; occasionally, infrequently, or they may never interact at all. What they will do, however, is identify each other through traditions they have inherited: a particular way of arranging the hair, a particular manner of dress, particular ornaments, particular visual art motifs and forms, as well as such things as a common language.

This cooperative behavior among individuals may also occur in other species. The Arabian babbler, a songbird that lives in Israel, behaves in ways that seem to be highly altruistic. Food is shared, as is care of nestlings and fledglings. Individuals who are heads of the hierarchy, rather than enjoying the fruits of their dominant status, are characterized by their generosity; they feed, care for, and protect the weaker and smaller babblers. Kin selection cannot explain this cooperation, as the babblers are not closely related. Nor can this cooperation be explained by reciprocal altruism, as this would lead to the prediction that babblers would cheat, taking the benefits of food sharing or nestling care without paying the costs. The Zahavis, who have studied these babblers for three decades, reject the possibility of group selection and argue instead that this seeming altruism is actually a type of handicap that gives the altruist higher status in the group and thus makes that altruist more attractive as a mate. Only the most fit individuals, those in the best possible condition, can afford to be altruists. Females looking for mates with good genes find them in the handicapped altruists.

Certain kinds of cooperation observed cross-culturally in humans that are associated with tribes and clans show some interesting parallels with this babbler behavior. Hospitality and generosity are given to clan "brothers and sisters, mothers and fathers," whether those individuals are acquaintances with whom one has regular interactions or they are strangers. The heads of the hierarchy, often the elders, are characterized by their greater responsibility; they are expected to protect those who are younger, weaker, and smaller. Food is shared, as are the responsibilities of childcare. Babbler behavior may provide no clues for understanding human behavior. Ancestry may be irrelevant to babblers (and the Zahavis surely would have noted if it were relevant); ancestry, however, has been of fundamental importance to humans. Both the handicap and ancestress hypothesis, however, rest upon a large investment made in individuals who are, based on kin-selection theory, only distantly related or are even nonkin.

The aim of this discussion is to point out that kinship behavior for humans is not, and for millennia probably has not been, limited to social behavior involving solely close kin (related more closely than second cousin). In Australia, kinship plays a central role. It is "the articulating force for all social interaction" (Berndt and Berndt 1977, 90). The same kin terms that are used for close kin are used for distant kin or metaphorical kin. There is "no terminological distinction made between close blood relatives and distant kin, so, for example, a person generally behaves toward all those women termed 'mother' in a similar patterned way," even though he will know which mother is his biological mother (genetrix), and he will have some mothers with whom he was raised while other mothers will be strangers to him. Every kin term "connotes a complex of ideal behaviors" (1978, 44).

## Concluding Remarks

The hypothesis proposed in this book focuses on mothers and their increasing maternal investment as catalysts for the evolution of modern human anatomy, physiology, and culture. This hypothesis can help account not only for traits such as the narrowing human pelvis and increasingly large brain but also adolescent subfecundity, concealed ovulation, an exterogestate fetus, menopause, allomothering, paternal care, and the long human life span. It can also help explain traditions, including those of visual art and moral systems, and the passionate behaviors associated with shared descent and traditions.

I have meandered through a number of topics in order to build the argument that the increased maternal investment in offspring was dependent upon enduring social relationships, which have high survival and mate-seeking costs. Humans regularly identify and cooperate with close kin; however, they also identify and cooperate with more distant kin—those actually sharing common ancestry. A later cultural invention, one associated with religion, was the use of a kinship metaphor to create a new ancestor (a mythical ancestor or a creator god) that encourages kinshiplike cooperation among a large category of individuals with distinct ancestors. Visual art's function, in either case, was to identify the codescendants and promote their cooperation.

To answer the questions raised at the beginning of this book, I argue that visual art, when used as our ancestors used it, has the characteristics associated with a good mother. Visual art encourages kin to cooperate with one another and make sacrifices for the more vulnerable without any assurance that there will be a commensurate return. Because competition is at the expense of social behavior, and a threat to vulnerable human offspring, traditions of visual art will limit such competition unless directed at nonkin— those not sharing ancestry. Just as art can be used to promote cooperation among kin, it can be used to promote animosity against nonkin.

The primary challenge with which this book leaves the reader, other than devising better tests for the ancestress and art hypothesis, is that it rests upon several fundamental but untested assumptions. One of these assumptions is that humans respond to and are influenced by a hierarchy, as the term has been defined in this book. Hierarchs are those whose influence comes from generosity and responsibility. In addition, this book also is built on the assumption that humans are copiers and are likely to copy those in hierarchical positions, those who are or who behave as if they are older relatives. Further, it assumes that kinship behaviors, including maternal, paternal, and sibling behaviors, are acquired through learning or copying that occurs during the developmental process. Finally, it assumes that maternal strategies, just as other kinship social behaviors, are not neutral. They can have multigenerational or dynastic effects.

# Glossary

*alloparent:* an individual other than the parent who helps care for a child (Wilson 1975). Allomothers are probably the earliest form of alloparenting; alloparenting was initially referred to as aunting behavior.

*altruism:* the act of promoting another's reproductive success at one's own reproductive expense (Alexander 1987). I use the terms *sacrifice* and *self-sacrifice* to refer to altruism.

*ancestors:* those who lived before us, from whom we inherited our genes and possibly our traditions. I do not use the term *ancestor,* as do some anthropologists (e.g., McAnanay 1995), to refer only to elite ancestors who are venerated and whose names and accomplishments have been passed down to some of us.

*ancestor worship or veneration:* a "religion in which deceased ancestors [are said to] actively influence the health and well-being of their descendants" (Moseley 1992, 53). The claim is also made that living descendants influence the ancestors through rituals, etc.

*ayllu:* a "kin collective" (set of co-descendants) that has a "founding ancestor and contains a number of lineages divided among two moieties" (Moseley 1992, 49). Membership is determined by birth. Members of an ayllu marry members of the same ayllu (marriage is endogamous). Within the moiety, marriage is exogamous, with one's spouse coming from the other moiety. Common ancestors give ayllus their ethnic identity; the *karacas'* (chiefs') influence resulted from close blood ties to ancestors (Moseley 1992).

*biological:* the nature of living matter. Thornhill and Palmer argue that the term *biological* refers to all aspects of living things (2000, 20). As everything biological, or living, is a result of interactions between genes and an environment, it is inaccurate to equate the term *biological* with the term *genetic* and to insist that biological implies inevitably.

*clans:* a "set of kin whose members believe themselves to be descended from a common ancestor" even though they may not be able to specify links back to the founder (Ember and Ember 1998, 374).

*competition:* the key to evolution, according to Darwin (1859); however, competition does not mean combat. Competition between ancestral mothers apparently occurred when some mothers began to move toward a greater K-strategy, even though this meant reducing the number of offspring they produced, while others did not increase their investment.

*culture:* once defined as "that complex whole which includes knowledge, belief, art, morals, law, custom, and any other capabilities and habits acquired by man as a member of a society" (Tylor 1958, 1); the essential conditions of culture are that it is not only learned and shared but also acquired from another individual and potentially transmittable to a third individual (Palmer and Steadman 1997).

*dynast:* an ancestress who lived and reproduced in the past and, due to the strategies she developed (or inherited) and transmitted to her offspring, promoted the reproductive and dynastic success of her descendants.

*enduring social relationships:* relationships that involve recurring cooperative interactions between two individuals, which may lead to the establishment of new relationships between the descendants of the respective individuals involved.

*environment of evolutionary adaptedness (EEA):* past environments; the environment in which our ancestors lived. Because our ancestors lived in a variety of environments, the EEA is a fiction, a composite (Wright 1994). Life in the EEA, however, is assumed to approximate the lifestyle found in contemporary foraging societies.

*fictive kin:* individuals who are, while not biological or blood kin, treated as if they were close kin and are referred to using kinship terms reserved for close kin. Also *metaphorical kin.*

*kinship:* identified through birth to a mother. Fathers, if identified, are identified through a prior sexual relationship with the mother. In order for kinship cooperation to occur, kin must be identified. Humans have an elaborate kinship system, extending kinship terminology metaphorically to distant kin and even to nonkin. Actual kinship behavior, however, refers to social behavior whose degree of altruism is correlated with genealogical distance (Palmer and Steadman 1997).

*lineage:* once defined as "a set of kin whose members trace descent from a common ancestor through known links" (Ember and Ember 1990:358). More commonly, it refers to the line of descendants going from ego back through a line of ancestors to the original ancestor.

*marriage:* a kinship relationship. In a tribe or clan involving an enduring, socially recognized, reproductive relationship formed between two codescendants of a common ancestor. Marriage is also an economic relationship in the sense that it involves a sexual division of labor.

*metaphorical kin:* See fictive kin.

*morals:* distinguished from laws in that morals are designed to promote enduring social relationships between kin (or metaphorical kin), while laws are designed to protect the rights of (certain) individuals in their interactions with strangers.

*noble:* having or showing high moral qualities or ideas; characterized by or characteristic of greatness of character. A person who behaves with nobility is self-sacrificing.

*reciprocity:* distinguished from altruism in that there is anticipation of a return on generosity. Because interaction is to the donor's benefit, reciprocal altruism is inherently a competitive situation with winners and losers.

*sacrifice:* an act that promotes another's reproduction at the expense of one's own (Alexander 1987). While real sacrifice may be extremely rare and an evolutionary mistake for the individual showing it, sacrifice may be a maternal strategy aimed at promoting a long-term, as opposed to a short-term, reproductive success.

*social behavior:* characterized by altruism or cooperation. Social behavior is contrasted with competitive behavior, which is antisocial. Asocial behavior does not involve interaction, altruism, or competition.

*totem:* a symbol of a clan or other category of individuals sharing common descent; e.g., a "plant or animal associated with a clan (sib) as a means of group identification" (Ember and Ember 1998, 382).

*tradition:* inheritable culture coming from the past, copied from one's own ancestors. Like other inheritable traits, traditions may influence the reproductive success of descendants and in so doing increase their own frequency in succeeding generations. Traditions may be subject to a form of natural selection. Unlike genes, traditions are potentially transmittable to all of our offspring and therefore can rapidly increase in frequency.

*traditional people (also traditional society):* those who possess a number of practices without which the transmission of traditions from one generation to the next would not occur: respect for the elders, the bearers of traditions, and the ancestors and their traditions. Rules in traditional societies encourage cooperation among codescendants, who share the traditions and frequently refer to one another using kinship terms. People who live a non-Western lifestyle but do not show these characteristics are not, by my definition, traditional people.

*tribe:* a large number of individuals who claim they share common ancestry and who may be subdivided into smaller divisions (e.g., moieties, clans, subsections), each of which shares its own (closer) ancestor in addition to the (more distant) tribal ancestor.

*visual art:* line, color, pattern, and/or form used by humans to modify an object or body solely to attract attention to that object or body.

# References


Abramova, Z. 1967. Paleolithic art in the USSR. *Arctic Anthropology* 4 (2): 1–179.

Adams, D., and D. Apostolos-Cappadona, eds. 1987. *Art as religious study*. New York: Crossroad.

Ady, C. 1993. Morals and manners of the Quattrocentro. In *Art and politics in Renaissance Italy*, ed. G. Holmes, 1–18. London: Oxford University Press.

Aiken, N. 1998. *The biological origins of art*. Westport, Conn.: Praeger.

Alcock, J. 1979. *Animal behavior*. Sunderland, Mass.: Sinauer.

———. 1998. *Animal behavior*. 2d ed. Sunderland, Mass.: Sinauer.

Alexander, R. 1974. The evolution of social behavior. *Annual Review of Ecology and Systematics* 5:325–383.

———. 1979. *Darwinism and human affairs*. Seattle: University of Washington Press.

———. 1987. *The biology of moral systems*. New York: Aldine de Gruyter.

———. 1990. *How did humans evolve? Reflections on the uniquely unique species*. Museum of Zoology, Special Publication No. 1. Ann Arbor: University of Michigan.

———. 2001. *Evolutionary selection and the nature of humanity*. Notre Dame: University of Notre Dame Press (forthcoming).

Alexander, R., J. Hoogland, R. Howard, K. Noonan, and P. Sherman. 1979. Sexual dimorphism and breeding systems in pinnipeds, ungulates, primates, and humans. In *Evolutionary biology and human social behavior: An anthropological perspective*, ed. N. Chagnon and W. Irons, 436–53. North Scituate, Mass.: Duxbury Press.

Alexander, R., and K. Noonan. 1979. Concealment of ovulation, parental care, and human social evolution. In *Evolutionary biology and human social behavior: An anthropological perspective*, ed. N. Changon and W. Irons, 402–35. North Scituate, Mass.: Duxbury Press.

Allen, M., and W. Lemmon. 1981. Orgasm in female primates. *American Journal of Primatology* 1:15–34.

Allport, G. 1954. *The nature of prejudice*. Cambridge, Mass.: Addison-Wesley.

Altmann, J. 1987. Life span aspects of reproduction and parental care in anthropoid primates. In *Parenting across the life span: Biosocial dimensions*, ed. J. Lancaster, J. Altmann, A. Rossi, and L. Sherrod, 15–30. New York: Aldine de Gruyter.

Altschuler, M. 1971. *The Cayapa: A study in legal behavior*. Ann Arbor: University Microfilms International.

Anderson, R. 1979. *Art in primitive society*. Englewood Cliffs, N.J.: Prentice-Hall.

———. 2000. *American muse: Anthropological excursions into art and aesthetics*. Upper Saddle River, N.J.: Prentice-Hall.

Anonymous. 1918. Editorial: Aesthetic propaganda. *The Art World and Arts and Decoration*.

Arensberg, C. 1968. *Family and Community in Ireland*. New Haven: Harvard University Press.

Arensberg, C., and S. Kimball. 1948. *Family and community in Ireland.* Cambridge, Mass.: Harvard University Press.

Aristotle. 1984. The Nicomachean ethics. In *Ethics: Selections from classical and contemporary writers,* 5th ed., ed. O. Johnson. New York: Holt, Rinehart and Winston.

Arriaza, B. 1995. *Beyond death: The Chinchorro mummies of ancient Chile.* Washington: Smithsonian Institution Press.

Ashby, H. 1915. *Infant mortality.* London: Arnold and Co.

Awiakta. M. 1993. *Selu: Seeking the corn-mother's wisdom.* Golden, Colo.: Fulcrum Publishing.

Baaren, Th. P. van. 1968. *Korwars and the Korwar style: Art and ancestor worship in north-west New Guinea.* The Hague: Mouton.

Bachofen, J. 1967. *Myth, religion, and Mother Right.* Bolingen Series. Princeton, N.J.: Princeton University Press.

Bahn, P. 1986. No sex please, we're Aurignacians. *Rock Art Research* 3:99–105.

———. 1991. Pleistocene images outside of Europe. *Proceedings of the Prehistoric Society* 57 (i): 99–102.

———. 1994. New advances in the field of Ice Age art. In *Origins of anatomically modern humans,* ed. M. Nitecki and D. Nitecki, 121–32. New York: Plenium.

———. 1995–96. New developments in Pleistocene Art. *Evolutionary Anthropology* 4 (6): 204–15.

Bahn, P., and J. Vertut. 1988. *Images of the Ice Age.* London: Windward.

Bailey, G., and J. Peoples. 1999. *Introduction to cultural anthropology.* Belmont, Calif.: West/Wadsworth.

Bakewell, L. 1995. Bellas artes and artes populares: The implications of difference in the Mexico City art world. In *Looking high and low: Art and cultural identity,* ed. B. Bright and L. Bakewell, 19–54. Tucson: University of Arizona Press.

Bakker, T., and B. Mundwiler. 1994. Female mate choice and male red coloration in a natural stickleback population. *Behavioral Ecology* 5:74–80.

Bandello, M. 1554. *Novelle.* Milan: Giovanni Silvestri. Printed by Nicholas and John Okes for James Becket, 1637.

Bar Josef, O. 1994. The Lower Paleolithic of the Near East. *Journal of World Prehistory* 8:211–65.

Barber, E. 1994. *Women's work: The first 20,000 years.* New York: W. W. Norton.

Barber, N. 1998. *Parenting roles: Styles and outcomes.* Commack, N.Y.: Nova Science.

———. 2000. *Why parents matter: Parental investment and child outcomes.* Westport, Conn.: Bergin and Garvey.

Barham, L. 2000. Prehistoric body paint. *Archaeological Institute of America* 53 (4): www.about_the_AIA/Press_Release.html. Summary by Elizabeth J. Himelfarb.

Barnard, A. 1999. Modern hunter-gatherers and early symbolic culture. In *The evolution of culture: An interdisciplinary view,* ed. R. Dunbar, C. Knight, and C. Power, 50–68. New Brunswick, N.J.: Rutgers University Press.

Baron, R., and D. Bryne. 1997. *Social psychology.* 8th ed. Boston: Allyn and Bacon.

Barrera, A. 1991. Mexican-American roadside crosses in Starr County. In *Hecho en Tejas: Texas-Mexican folk arts and crafts,* ed. J. Graham, 278–308. Denton: University of North Texas Press.

Barrera-Vásquez, A., et al. 1980. *Diccionario Maya Cordemex*. Mérida, Yucatán: Ediciones Cordemex.

Barrett, S. 1925. *Indian notes and monographs. Cayapa Indians of Ecuador, part I and II*. New York: Heye Foundation.

Barrow, T. 1979. *The art of Tahiti and the neighbouring Society, Austral and Cook Islands*. London: Thames and Hudson.

Bateson, G. 1971. Style, grace, and information in primitive art. In *Steps to an ecology of mind*, ed. G. Bateson. New York: Ballantine.

Bateson, P. 1980. Optimal outbreeding and the development of sexual preferences in Japanese quail. *Zeitschrift für Tierpschylogie*. 53:231–44.

Baumann, H. 1935. *Lunhda*. Berlin: Wörfel Verlag.

Beardsley, G. 1958. *Aesthetics: Problems in the philosophy of criticism*. New York: Harper and Row.

Bell, C. 1958. *Art*. New York: Capricorn.

Bell, D. 1983. *Daughters of the dreaming*. Sydney: George Allen and Unwin Australia.

Berner, R. T. 1992. *Parents whose parents were divorced*. New York: Haworth Press.

Berenson, B. 1948. *Aesthetics and history*. Garden City, N.Y.: Doubleday and Co.

Bernsdorf, L., and R. Thornhill. 1979. The evolution of monogamy and concealed ovulation in humans. *Journal of Social and Biological Structures* 2:95–106.

Berndt, R., ed. 1964. *Australian Aboriginal art*. Sydney: Ure Smith.

Berndt, R., and C. Berndt. 1977. *The world of the first Australians*. Sydney: Ure Smith.

Bereczkel, T. 1993. R-selected reproductive strategies among Hungarian gypsies: A preliminary analysis. *Ethology and Sociobiology* 14:71–78.

Bersani, L., and U. Dutoit. 1985. *The forms of violence: Narrative in Assyrian art and modern culture*. New York: Schocken Books.

Biebuyck, D. 1973. *Lega culture: Art, initiation, and moral philosophy among a central African people*. Berkeley: University of California Press.

Biesele, M., and N. Howell. 1981. "The old people give you life": Aging among !Kung hunter-gathers. In *Other ways of growing old*, ed. P. Amoss and S. Harrell. Stanford: Stanford University Press.

Biggs, P. 1989. *Art, death and social order: The mortuary arts of pre-conquest central Panama*. Oxford: B. A. R. International Series.

Binford, L. R. 1971. Mortuary practices: Their study and their potential. In *Approaches to the social dimensions of mortuary practices*, ed. J. Brown, 6–29. Washington, D.C.: Society for American Archaeologists. Issued as *American Antiquity* 35 (July 1971). Memoirs of the Society for American Archaeology, No. 25.

Bird, I. 1984. *Unbeaten tracks in Japan: An account of travels in the interior, including visits to the Aborigines of Yezo and the shrine of Nikko*. Tokyo: Charles E. Tuttle.

Birkhead, T., and A. Møller. 1996. Monogamy and sperm competition in birds. In *Partnerships in birds: The study of monogamy*, ed. J. Black, 232–343. Oxford: Oxford University Press.

Bland, K. 2000. *The artless Jew: Medieval and modern affirmations and denials of the visual*. Princeton, N.J.: Princeton University Press.

Blake, F. 1955. *Primitive art*. New York: Dover Publications.

Blake, J. 1955. *Family instability and reproductive behavior in Jamaica.* Milbank Memorial Fund Current Research in Human Fertility. New York: Milband.

Blier, S. 1998. *The royal arts of Africa.* New York: Prentice-Hall.

Blocker, H. 1979. *Philosophy of art.* New York: Charles Scribner's Sons.

Blos, V. 2001. *Decoraciones dentales entre los antiguos Mayas.* Mexico, D.F.: Ediciones Euroamericanas. Instituto Nacional de Antropología e Historia, Cordoba.

Blurton-Jones, N., K. Hawkes, and J. O'Connell. 1997. Why do Hadza children forage? In *Uniting psychology and biology: Integrative perspectives on human development,* ed. N. Segal, G. Weisfeld, and C. Weisfeld, 279–313. Washington, D.C.: American Psychological Association.

Boas, F. 1955. *Primitive art.* New York: W. W. Norton.

Boesch, C. 1990. First hunters of the forest. *New Scientist* 125:605–13.

———. 1993. Aspects of transmission of tool-use in wild chimpanzees. In *Tools, language, and cognition in human evolution,* ed. K. Gibson and T. Ingold. Cambridge, U.K.: Cambridge University Press.

Boesch, C., and H. Boesch. 1989. Hunting behavior of wild chimpanzees in the Taï National Park. *American Journal of Physical Anthropology* 78:547–74.

Bongaarts, J. 1980. Malnutrition and fecundity. *Studies in Family Planning* 11:401–6.

Boone, J., and K. Kessler. 1999. More status or more children? Social status, fertility reduction, and long-term fitness. *Evolution and Human Behavior* 20:257–77.

Borbolla, D. 1940. Types of tooth mutilation found in Mexico. *American Journal of Physical Anthropology* 25:631–45.

Bordes, F. 1961. Mousterian cultures in France. *Science* (September 22): 803–10.

Bordes, F. 1972. *A tale of two caves.* New York: Harper and Row.

Boyce, W., H. Shafer, H. Harrison et al. 1986. Social and cultural factors in pregnancy complications among Navajo women. *American Journal of Epidemiology* 124:242–53.

Boyd, R., and P. Richerson. 1985. *Culture and the evolutionary process.* Chicago: University of Chicago Press.

———. 1990. Culture and cooperation. In *Beyond self-interest,* ed. J. Mansbridge. Chicago: University of Chicago Press.

Braun, B., ed. 1995. *Arts of the Amazon.* London: Thames and Hudson.

Briffault, R. 1931. *The mothers: The matriarchal theory of social origins.* New York: Macmillan Co.

Broman Morales, V., and R. Braidwood. 1991. Shadows of doubt in identifying female images: A reply to Kehoe. *Antiquity* 65:914–15.

Brookfield, H., and P. Brown. 1963. *Struggle for land: Agriculture and group territories among the Chumbu of the New Guinea highlands.* New York: Oxford University Press.

Brothwell, D. 1976. Towards a working definition of visual art. In *Beyond aesthetics,* ed. D. Brothwell, 10–17. London: Thames and Hudson.

Brown, J. 1972. Plato's Republic as an early study of media bias and a charter for prosaic education. In *American Anthropologist* 75 (3): 672–75.

Bruner, E. M. 1961. Differential change in the culture of the Mandan from 1250–

1953. In *Perspectives in American Indian Culture Change,* ed E. H. Spicer. Chicago: University of Chicago Press.

Bryce, R. 1991. Support in pregnancy. *International Journal of Technology Assessment in Health Care* 7:478–84.

Burbank, V. 1998. Women's intra-gender relationships and "disciplinary aggression" in an Australian Aboriginal community. In *Women among women: Anthropological perspectives on female age hierarchies,* ed. J. Dickerson-Putman and J. Brown, 68–77. Urbana: University of Illinois Press.

Burch, R., and G. Gallup. 2000. Perceptions of paternal resemblance predict family violence. *Evolution and Human Behavior* 21:429–35.

Burley, N. 1979. The evolution of concealed ovulation. *American Naturalist* 114: 835–58.

Burnett, J. 1839. Captain Abraham McClellan's Company, 2d Regiment, 2d Brigade, Mounted Infantry, Cherokee Indian Removal, 1838–39, at www.bower source.com/cherokee/burnett.html.

Buss, D. 1998. The psychology of human mate selection: Exploring the complexity of the strategic repertoire. In *Handbook of evolutionary psychology,* ed. C. Crawford and D. Krebs, 405–30. Mahwah, N.J.: Lawrence Erlbaum Associates.

Byers, B., and D. Kroosdma. 1992. Development of two song categories by chestnut-sided warblers. *Animal Behavior* 44:799–810.

Cabanne, P. 1971. *Dialogues with Marcel Duchamp.* Trans. from the French by Ron Padgett. New York: Viking Press.

Calhoun, J. 1990. Plight of the Ik and Kaiadilt is seen as a chilling possible end for man. In *Anthropology: Contemporary perspectives,* ed. P. Whitten and D. Hunter, 208–13. Glenview, Ill.: Scott, Foresman and Co.

Cantú, N. 1991. Costume as cultural resistance and affirmation. In *Hecho en Tejas: Texas-Mexican folk arts and crafts,* ed. J. Graham, 117–45. Denton: University of North Texas Press.

Campbell, C. 1960. The susceptibility of mother mice and pregnant mice to the virus of foot and mouth disease. *Journal of Immunology* 84:469–74.

Campbell, D. 1975. On the conflicts between biological and social evolution and between psychology and moral tradition. *American Psychologist* 30:1103–26.

Campbell, T. 1925. *Dentition and palette of the Australian Aboriginal.* Sydney: Hassel Press.

Cárdenas, F. In press. Momias, santuarios y ofrendas: El contexto ritual de la momificación en el altiplano central de los Andes Colombianos. In *Memorias del Congreso Internacional de Estudios Sobre Momias.* Santa Cruz, Spain: Tenerife.

Carmack, R. 1981. *The Quiché Mayas of Utatlan.* Norman: University of Oklahoma.

Caro, T 1994. *Cheetahs of the Serengeti Plains: Group living in an asocial species.* Chicago: University of Chicago Press.

Castro, F., K. Coe, S. Gutierres, D. Saenz. 1996. Designing health promotion programs for Latinas. In *Health psychology of special populations,* ed. P. Kato and A. Mann, 319–45. New York: Plenium.

Castro, F., K. Coe, and M. Harmon. 1996. The effect of ethnic/racial matches between provider and patient on the use of health services by Hispanics and

African Americans. In *Achieving equitable access,* ed. M. Lillie-Blanton, W. Leigh, and A. Alfaro-Correa, 7–27. Washington, D.C.: Joint Center for Political and Economic Studies.

Cattell, M. 1990. "Nowadays it isn't easy to advise the young": Grandmothers and granddaughters among Abaluyia of Kenya. In *Women among women: Anthropological perspectives on female age hierarchies,* ed. J. Dickerson-Putman and J. Brown, 30–50. Urbana: University of Illinois Press.

Cavalli-Sflorza, L., and M. Feldman. 1981. *Cultural transmission and evolution.* Princeton, N.J.: Princeton University Press.

Chadwick, W. 1990. *Women, art, and society.* London: Thames and Hudson.

Chagnon, N. 1977. *Yanamamö: The fierce people.* New York: Holt, Rinehart and Winston.

Chastel, A. 1983. *The sack of Rome, 1527.* A. W. Mellon Series in the Fine Arts. Princeton, N.J.: Princeton University Press.

Cheney, D., and R. Seyfarth. 1990. *How monkeys see the world.* Chicago: University of Chicago Press.

Chesser, E. 1956. *The sexual, marital and family relationships of the English woman.* Great Britain: Hutchinson's Medical Publishers.

Cloak, F. 1975. Is a cultural ethology possible? *Human Ecology* 3:161–82.

Clottes, J. 1997. New laboratory techniques and their impact on Paleolithic cave art. In *Beyond art: Pleistocene art and symbol,* ed. M. Conkey, O. Soffer, D. Stratmann, and N. Jablonski, 37–52. San Francisco: California Academy of Sciences.

Clough, G. 1965. Variability in wild meadow voles under various conditions of population density, season, and reproductive activity. *Ecology* 46:119–34.

Coe, K. 1972. Art: The replicable unit. Presentation given to the American Anthropological Association, Philadelphia, December.

————. 1992. Art: The replicable unit. An Inquiry into the origin of art as a social behavior. *Journal of Social and Evolutionary Systems* 15:217–34.

————. 1995. Lectures from the ancestresses: The moral system and art of the Chachi of lowland Ecuador. Ph.D. diss., Arizona State University.

Coe, K., and L. Steadman. 1995. The human breast and the ancestral reproductive cycle: A preliminary inquiry into breast cancer etiology. *Human Nature* 6 (3): 197–220.

Coe, K., and C. Keller. 1996. Health protective behaviors of young African American women. Should we be using a kinship model to teach health behaviors? *Journal of Human Ecology. Special Issue No. 4: The Family as an Environment for Human Evolution,* 57–66.

Cole H. 1989. *Icons and power: The arts of Africa.* Washington D.C.: Smithsonian Press.

Collins, D. 1976. *The human revolution: From ape to artist.* New York: E. P. Dutton.

Colwell, M., and L. Oring. 1989. Extra-pair mating in the spotted sandpiper: A female mate acquisition tactic. *Animal Behavior* 38:675–84.

Conkey, M. 1983. On the origins of Paleolithic art: A review and some critical thoughts. In *The Mousterian legacy human biocultural change in the Upper Pleistocene,* ed. E. Trinkaus, 201–27. Oxford: BAR International Series, British Archaeological Reports.

———. 1997. Context in the interpretive process. In *Beyond art: Pleistocene art and symbol,* ed. M. Conkey, O. Soffer, D. Stratmann, and N. Jablonski, 343–67. San Francisco: California Academy of Sciences.

Contenau, G. 1954. *Everyday life in Babylon and Assyria.* Trans. K. and A. Maxwell-Hyslop. London: E. Arnold.

Cory, H. 1956. *African figurines: Their ceremonial use in puberty rites in Tanganyika.* New York: Grover.

Coulange de, F. 1955. *The ancient city.* Garden City, N.Y.: Doubleday.

Crystal, E. 1994. Rape of the ancestors. In *Fragile Traditions,* ed. P. Tylor, 29–41. Honolulu: University of Hawaii Press.

Cunningham, A. 1995. Breastfeeding: Adaptive behavior for child health and longevity. In *Breastfeeding: Biocultural perspectives,* ed. P. Stuart-Macadem and K. Dettwyler, 243–64. New York: Aldine de Gruyter.

D'Acevedo, W. 1993. Mask makers and myth in western Liberia. In *Arts of Africa, Oceania, and the Americas: Selected Readings,* ed. J. Berlo and L. Wilson, 111–32. Englewood Cliffs, N.J.: Prentice-Hall

Daly, M. 1990. Evolutionary theory and parental motives. In *Mammalian parenting,* ed. N. Krasnegor and R. Bridges, 25–39. New York: Oxford University Press.

Daly, M., and Wilson, M. 1985. Child abuse and other risks of not living with both parents. *Ethology and Sociobiology* 6:197–210.

———. 1988a. *Homicide.* Hawthorne, N.Y.: Aldine de Gruyter.

———. 1998b. The evolutionary social psychology of family violence. In *Handbook of evolutionary psychology,* ed. C. Crawford and D. Krebs, 431–56. Mahwah, N.J.: Lawrence Erlbaum Associates.

Danto, A. 1964. The artworld. *Journal of Philosophy* 6:22–25.

Darwin, C. 1859. *On the origin of species.* Cambridge, Mass.: Harvard University Press.

———. 1871. *The descent of man and selection in relation to sex.* 2 vols. London: John Murray.

———. 1887. *The autobiography of Charles Darwin and selected letters.* 3 vols. London: John Murray.

Davidson, I. 1997. The power of pictures. In *Beyond art: Pleistocene art and symbol,* ed. M. Conkey, O. Soffer, D. Stratmann, and N. Jablonski, 125–60. San Francisco: California Academy of Sciences.

Dawkins, R. 1976. *The selfish gene.* Oxford: Oxford University Press.

———. 1982. *The extended phenotype.* Oxford: Oxford University Press.

Dawkins, R., and J. Krebs. 1978. Animal signals: Information or manipulation? Introduction to *Behavioral Ecology,* ed. J. R. Krebs and N. B. Davies, 282–330. Oxford: Blackwell Scientific.

De la Croix, H., and R. Tansey. 1980. *Gardner's art through the ages.* 7th ed. New York: Harcourt Brace Jovanovich.

De Lange, M. 1961. Dolls for the promotion of fertility as used by some of the Nguni tribes and the Basuto. *Annals of the Cape Provincial Museums* 1:86–101.

De Waal, F. 1996. Survival of the kindest: A simian Samaritan shows nature's true heart. *New York Times,* August 22.

Deacon, H. 1989. Late Pleistocene paleontology and archaeology in the Southern Cape, South Africa. In *The human revolution: Behavioral and biological perspectives on the origins of modern humans,* ed. P. Mellars and C. Stringer, 447–564. Princeton: Princeton University Press.

Delporte, H. 1979. *L'Image de la Femme dans l'Art Préhistorique.* Paris: Picard.

———. 1990. *L'image des Animaux dans l'Art Préhistorique.* Paris: Picard.

Dennett, H. 1975. *Mak Bilong Sepik: A selection of designs and paintings from the Sepik River.* Wewak, New Guinea: Wirui Press.

Dettwyler, K. 1995a. The hominid blueprint for the natural age of weaning in modern human populations. In *Breastfeeding: Biocultural perspectives,* ed. P. Stuart-Macadem and K. Dettwyler, 39–74. New York: Aldine de Gruyter.

———. 1995b. Beauty and the breast: The cultural context of breastfeeding in the United States. In *Breastfeeding: Biocultural perspectives,* ed. P. Stuart-Macadem and K. Dettwyler, 167–216. New York: Aldine de Gruyter.

———. 2000. More than nutrition: Breastfeeding in urban Mali. In *Nutritional anthropology: Biocultural perspectives on food and nutrition,* ed. A. Goodman, D. Dufour, and G. Pelto, 312–20. Mountain View, Calif.: Mayfield Publishing Co.

Dewey, J. 1958. *Art as experience.* New York: G. Putnam's Sons.

Diamond, J. 1991. *The rise and fall of the third chimpanzee.* London: Radius.

———. 1992. *The third chimpanzee.* New York: Harper Collins.

———. 1997. Changing sex is hard to do. *Science* 275:1745.

Dickerson-Putman, J., and J. Brown. 1998. *Women among women: Anthropological perspectives on female age hierarchies.* Urbana: University of Illinois Press.

Dickie, G. 1971. *Aesthetics: An introduction.* Indianapolis: Pegasus.

———. 1974. *Art and the aesthetic: An institutional analysis.* Ithaca: Cornell University.

———. 1997. Art: Function of procedure—nature or culture? *Journal of Aesthetics and Art Criticism* 55 (1): 19–28.

Dickie, G., and R. Sclafani. 1977. *Aesthetics: A critical anthology.* New York: St. Martin's.

Dingwell, E. 1931. *Artificial cranial deformation: A contribution to the study of ethnic mutilations.* London: John Bale Sons and Danielsson.

Dissanayake, E. 1992. *Homo aestheticus.* New York: Free Press.

———. 2000. *Art and intimacy: How the arts began.* Seattle: University of Washington Press.

Driver, G., and J. Miller. 1935. *The Assyrian laws.* Oxford: Clarendon Press.

Dunbar, R. 1987. Demography and reproduction. In *Primate societies,* ed. B. Smuts, D. Cheney, R. Sefarth, R. Wrangham, and T. Struhsaker, 240–49. Chicago: University of Chicago Press.

———. 1993. Coevolution of neocortical size, group size, and language in humans. *Behavioral and Brain Sciences* 16:681–735.

———. 1995. The mating system of callitrichid primates: I. Conditions for the coevolution of pair bonding and twinning. *Animal Behavior* 50:1057–70.

Dunbar, R., C. Knight, and C. Power. 1999. An evolutionary approach to human culture. In *The evolution of culture: An interdisciplinary view,* ed. R. Dunbar, C. Knight, and C. Power, 1–11. New Brunswick, N.J.: Rutgers University Press.

Duras, M. 1984. *The lover.* Trans. P. Bray. New York: Harper Perennial, 1992.

Durham, W. 1976. The adaptive significance of cultural behavior. *Human Ecology* 4:89–121.

Duviols, P. 1978. Un symbolisme andin du double: La lithomorphose de l'ancetre. *Actes du XLII Congres International Des Américanistes* 4:359–64.

Edel, M., and A. Edel. 1959. *Anthropology and ethics.* Springfield, Ill.: Charles C. Thomas.

Ehrenberg, M. 2001. The role of women in human evolution. In *Gender in cross-cultural perspective,* ed. C. Brettell and C. Sargent, 17–22. Upper Saddle River, N.J.: Prentice-Hall.

El Najjar, M., and G. Dawson. 1977. The effect of artificial cranial deformation on the incidence of wormian bones in the lambdoidal suture. *American Journal of Physical Anthropology* 461:155–60.

Elkin, A. 1933. Studies in Australian totemism: Sub-section, section, and Moiety totemism. *Oceania* 4 (1): 65–90

———. 1953. Murngin kinship re-examined and remarks on some generalizations. *American Anthropologist* 55:412–19.

———. 1964. *The Australian Aborigines.* Garden City, N.Y.: Doubleday.

Elkin, A., and R. Berndt. 1950. *Art in Arnhem Land.* Chicago: University of Chicago Press.

Ellison, P. 1982. Skeletal growth, fatness, and menarcheal age: A comparison of two hypotheses. *Human Biology* 54:269–81.

Ember, C., and M. Ember. 1990. *Cultural anthropology.* 6th ed. Englewood Cliffs, N.J.: Prentice-Hall.

———. 1998. *Anthropology: A brief introduction.*3d ed. Upper Saddle River, N.J.: Prentice-Hall.

Ergang, R. 1967. *The renaissance.* Princeton, N.J.: D. Van Nostrand Co.

Evans-Pritchard, E. 1940. *The Nuer.* London: Oxford University Press.

———. 1949. *The Sanusi of Cyrenaica.* Oxford: Clarendon Press.

———. 1965. *Nuer religion.* Oxford: Clarendon Press.

Faber, P., and L. Glasgow. 1968. Factors modifying host resistance to viral infections: II. Enhanced susceptibility of mice to encephalomycarditis virus infection during pregnancy. *American Journal of Pathology* 53:463–78.

Faber, L., and H-J. Martin. 1976. *The coming of the book: The impact of printing, 1450–1800.* Trans. David Gerard. London: NLB.

Fagen, B. 1990. *The journey from Eden: The peopling of our world.* London: Thames and Hudson.

Fairbanks, L., and M. McGuire. 1986. Determinants of fecundity and reproductive success in captive vervet monkeys. *American Journal of Primatology* 7:27–38.

Faris, J. 1972. *Nuba personal art.* London: Duckworth.

Fastlicht, S. 1971. *La odontologia en Mexico prehispanico.* Mexico, D.F.: Derechos Reservados.

Feeley-Harnik, G. 1985. Issues in divine kinship. *Annual Review of Anthropology* 14: 273–313.

Fernandez, J. 1972a. Tabernanthe Iboga: Narcotics ecstasies and the work of the ancestors. In *Flesh of the gods: The ritual use of hallucinogens,* ed. P. Furst, 237–60. London: George Allen and Unwin.

———. 1972b. Persistence and performance. *Daedalis* (winter): 39–60.

Feuchtwang, S. 1974. *An anthropological analysis of Chinese geomancy.* Taipei: Vithagna, Southern Materials Center.

Feyisetan, B., and Bankole, A. 1991. Mate selection and fertility in urban Nigeria. *Journal of Comparative Family Studies* 22:273–92.

Findlay, D. 1974. The role of suckling in lactation. In *Lactogenic hormones, fetal nutrition, and lactation,* J. Josimovich, M. Reynolds, and E. Cobo, 453–77. New York: John Wiley.

Fisher, H. 1982. *The sex contract: The evolution of human behavior.* New York: Morrow.

Fleming, W. 1974. *Arts and ideas.* New York: Holt, Rinehart and Winston.

Flinn, M. 1997. Family environment, stress, and health during childhood. In *Hormones, health, and behavior,* ed. C. Panter–Brick and C. Worthman, 105–38. Cambridge, U.K.: Cambridge University Press.

Flinn, M., and B. England. 1995. Childhood stress and family environment. *Current Anthropology* 36:854–86.

Flinn, M., and B. Low. 1986. Resource distribution, social competition, and mating patterns in human societies. In *Ecological aspects of social evolution: Birds and mammals,* ed. D. Rubenstein and R. Wrangham, 217–43. Princeton, N.J.: Princeton University Press.

Flood, J. 1983. *Archaeology of the dreamtime. The story of prehistoric Australia and its people.* Rev. ed. New Haven, Conn.: Yale University Press.

Foley, R., and M. Lahr. 1997. Mode 3 technologies and the evolution of modern humans. *Cambridge Archaeological Journal* 7 (1): 3–36.

Foley, R., and P. Lee. 1989. Finite social space, evolutionary pathways, and reconstructing hominid behavior. *Science* 243:901–6.

Fong, W. 1984. *Images of the mind.* Princeton, N.J.: Art Museum.

Forge, A., ed. 1973. *Primitive art and society.* London: Oxford University Press, Wenner-Gren Foundation.

Fortes, M. 1969. *Kinship and the social order.* London: Routledge.

Fox, R. 1993. *Reproduction and succession: Studies in anthropology, law, and society.* London: Transaction Publishers.

Francastel, P. 1967. *Medieval painting.* New York: Laurel.

Franciscan Fathers. 1910. *An ethnologic dictionary of the Navajo language.* St. Michaels, Ariz.: St. Michael's Press.

Freedberg, D. 1989. *The power of images: Studies in the history and theory of response.* Chicago : University of Chicago Press.

Freedman, M. 1966. *Chinese lineage and society: Fukien and Kwangtung.* London School of Economic Monographs on Social Anthropology, No. 33. New York: Athlone Press.

Freeman, D., and A. Barnes. 1959. Deaths from Asian influenza associated with pregnancy. *American Journal of Obstetrics and Gynecology* 78:1172–75.

Frisch, R., and J. MacArthur. 1974. Menstrual cycles: Fatness as a determinant of maximum weight for height necessary for their maintenance or onset. *Science* 185:949–51.

Frischknecht, M. 1993. The breeding colouration of male three-spined sticklebacks (*Gasterosteus aculeatus*) as an indicator of energy investment in vigour. *Evolutionary Ecology* 7:348–56.

Galef, B. 1983. Costs and benefits of mammalian reproduction. In *Symbiosis in parent-offspring interactions,* ed. L. Rosenblum and H. Moltz, 249–78. New York: Plenum Press.

Gallin, R. 1998. The intersection of class and GE: Mother-in-law/daughter-in-law relations in rural Taiwan. In *Women among women: Anthropological perspectives on female age hierarchies,* ed. J. Dickerson-Putman and J. Brown, 1–14. Urbana: University of Illinois Press.

Gamble, C. 1982. Interaction and alliance in Paleolithic society. *Man,* n.s., 71:92–107.

Gambutis, M. 1991. *The civilization of the goddess.* San Francisco: Harper.

Gardner, H. 1998. Do parents count? Review of J. R. Harris, *The nurture assumption: Why do children turn out the way they do? New York Review of Books 45* (17): 22.

Gardner, R., B. Gardner, B. Chiarelli, and F. Plooij. 1994. *The ethological roots of culture.* Dordrecht: Kluwer Academic Publishers.

Geary, D. 1998. *Male, female: The evolution of human sex differences.* Washington, D.C.: American Psychological Association.

———. 2000. Evolution and proximate expression of human paternal investment. *Psychological Bulletin* 126:55–77.

Gerbrands, A. 1967. *Wow-Ipits: Eight Asmat woodcarvers of New Guinea.* The Hague: Mouton.

———. 1971. Art as an element of culture in Africa. In *Anthropology and art: Readings in cross-cultural aesthetics,* ed. C. Otten, 366–82. Austin: University of Texas Press.

Ghiglieri, M. 1987. Sociobiology of the great apes and the hominid ancestor. *Journal of Human Evolution* 16:319–57.

Gilles, H., J. Lawson, M. Sibelas, A. Voller, and N. Allan. 1969. Malaria, anemia, and pregnancy. *Annals of Tropical Medicine and Parasitology* 63:245–63.

Glasse, R. 1965. The Huli of the southern Highlands. In *Gods, ghosts, and men in Melanesia,* ed. P. Lawrence and M. Meggitt, 27–49. New York: Oxford University Press.

Glassman, D., A. Coelho, K. Carey, and C. Bramblett, 1984. Weight growth in savannah baboons: A longitudinal study from birth to adulthood. *Growth* 48:425–33.

Glaze, A. 1993. Women power and art in a Senufo village. In *Arts of Africa, Oceania, and the Americas: Selected readings,* ed. J. Berlo and L. Wilson, 217–30. Englewood Cliffs, N.J.: Prentice-Hall.

Goldwater, R., and M. Treves, eds. 1945. *Artists on art: From the fourteenth to the twentieth century.* New York: Pantheon Books.

186 **REFERENCES**

Gombrich, E. 1972. *Art and illusion: A study in the psychology of pictorial representation.* Bollingen Series 35 (5). Princeton, N.J.: Published for Bollingen Foundation by Princeton University Press.

———. 1989. *The story of art.* Englewood Cliffs, N.J.: Prentice-Hall.

Goodale, J. 1971. *Tiwi wives: A study of the women of Melville Island, North Australia.* Seattle: University of Washington Press.

Goodall, J. 1970. Tool-using in primates and other vertebrates. *Advances in the Study of Behavior* 3:195–249.

———. 1971. *In the shadow of man.* Boston: Houghton Mifflin.

———. 1973. Cultural elements in a chimpanzee community. In *Precultural primate behaviour.* Vol. 1. Ed. E. Menzel. Basel: Kareger, Fourth IPC Symposium Proceedings.

———. 1984. The nature of the mother-child bond and the influence of the family on the social development of free-living chimpanzees. In *The growing child in family and society,* ed. N. Kobayashi and T. B. Brazelton, 47–66. Tokyo: University of Tokyo Press.

———. 1986. *The chimpanzees of Gombe: Patterns of behavior.* Cambridge, Mass.: Harvard University Press.

———. 1999. *Reason for hope: A spiritual journey.* New York: Time Warner.

Goody, J. 1962. *Death, property and the ancestors: A study of the mortuary customs of the LoDagaa of West Africa.* London: Tavistock.

Goody, J., and S. Tambiah. 1973. *Bridewealth and dowry.* Cambridge, U.K.: University of Cambridge Press.

Gootleib, G. 1992. *Individual development and evolution: The genesis of novel behavior.* New York: Oxford University Press.

Gose, P. 1992. Segmentary state formation and ritual control of water under the Incas. *Comparative Study of Society and History* 35:480–514.

Graham, J., ed. *Hecho en Tejas: Texas-Mexican folk arts and crafts.* Denton: University of North Texas Press, 1991.

Gramercy Great Masters. 1996. *Sandro Botticelli.* New York: Avenet.

Gregoriev, G. 1996. Le travail de l'ivoire sur la plaine Russe: le cas d'Avdeevo. In *Le Traveil et l'usage de l'Ivoire au Paléolithique Supéireur,* ed. M. Menu, P. Walker, and F. Widemann. Ravello. Italy: Centre Universitaire Européen pour les Biens Culturels.

Gribbon, J., and M. Gribbin. 1988. *The one per cent advantage: The sociobiology of being human.* New York: Basil Blackwell.

Grieder, T. 1982. *Origins of pre-Columbian art.* Austin: University of Texas Press.

Griffen, D. 1984. Animal thinking. *American Scientist* 72:456–64.

Griffin-Pierce, T. 1992. *Earth is my mother, sky is my father: Space, time, and astronomy in Navajo sandpainting.* Albuquerque: University of New Mexico.

Grün, R., and C. Stringer. 1991. Electron spin resonance dating and the evolution of modern humans. *Archaeometry* 33:153–99.

Guemple, L. 1995. Gender in Inuit society. In *Women and Power in Native North America,* ed. L. Klein and L. Ackerman, 17–27. Norman: University of Oklahoma.

Haeberlin, H. 1918. Principles of esthetic form in the art of the North Pacific Coast. *American Anthropologist* 20 (3): 258–64.

Haglund, L. 1976. *An archaeological analysis of the Broadbeach Aboriginal burial ground.* St. Lucia, Queensland: University of Queensland Press.

Haile, B. 1947. *Navaho sacrificial figurines.* Chicago: University of Chicago Press.

Haldane, J. 1932. *The causes of evolution.* New York: Harper and Brother.

———. 1955. Population genetics. *New Biology* 18:34–51.

Hall, M. 1991. Health of pregnant women. *British Medical Journal* 303:460–62.

Halliday, T. 1983. The study of mate choice. In *Mate choice,* ed. P. Bateson, 3–32. Cambridge, U.K.: Cambridge University Press.

Halverson, J. 1987. Art for art's sake in the Paleolithic. *Current Anthropology* 28:63–87.

Hambly, W. 1925. *The history of tattooing and its significance.* London: H. F. and G. Witherby.

Hamburg, D. 1952. Relevance of recent evolutionary changes to human stress biology. In *Social life of early man,* ed. S. Washburn, 278–88. Chicago: Aldine.

———. 1963. Emotions in the perspective of human evolution. In *Expression of the emotions in man,* ed. P. Knapp, 300–317. New York: International Universities.

Hames, R. 1989. The allocation of parental care among the Yekwana. In *Human reproductive behavior: A Darwinian perspective,* ed. L. Betzig, M. Borgerhofff-Mulder, and P. Turke, 237–52. Cambridge, U.K.: Cambridge University Press.

Hamilton, A. 1970. The role of women in Aboriginal marriage arrangements. In *Women's role in Aboriginal society,* ed. F. Gale, 17–20. Canberra: Australian Institute of Aboriginal Studies.

———. 1981. *Nature and nurture: Aboriginal child-rearing in north-central Arnhem Land.* Canberra: Australian Institute of Aboriginal Studies.

Hamilton, W. 1964. The genetical evolution of social behavior, II. *Journal of Theoretical Biology* 7:17–52.

Haraway, D. 1989. *Primate visions.* New York: Routledge.

Hart, C., and A. Pilling. 1960. *The Tiwi of North Australia.* New York: Holt, Rinehart and Winston.

Hart Hansen, J., and J. Nordqvis. 1996. The mummy find from Qilakitsoq in northwest Greenland. In *The man in the ice.* Vol. 3, *Human mummies: A global survey of their status and the techniques of conservation,* ed. K. Spindler, H. Wilfing, E. Rastbichler-Zissernig, D. zur Nedden, H. Nothdurfter, 107–21. New York: Springer Wien.

Harrenkohl, L. 1986. Prenatal stress disrupts reproductive behavior and physiology in offspring. *Annals of the New York Academy of Sciences* 74:121–28.

Hauser, A. 1959. *The social history of art: Prehistoric, ancient-Oriental, Greece, and Rome, Middle Ages.* 2 vols. New York: Vintage Books.

Havelock, E. 1963. *Preface to Plato.* Cambridge, Mass.: Belknap Press of Harvard University.

Hawkes, K. 1977. Co-operation in binumarien: Evidence for Sahlin's model. *Man* 12:459–83.

Hawkes, K., H. Kaplan, K. Hill, and A. Hurtado. 1997. Aché at the settlement: Contrasts between farming and foraging. *Human Ecology* 15:133–61.

Heider, K. 1979. *Grand Valley Dani*. New York: Holt, Rinehart and Winston.

Heizer, R., and K. Nissen. 1977. Prehistoric rock art of Nevada and California. *Sonderdruck aus IPEK, Jarbuch für Prehistorische and Ethnographische Kunst*. Vol. 24. Berlin: Walter de Gruyter.

Held, V. 1987. Non-contractual society. In *Science, morality and feminist theory*, ed. M. Hanen and K. Nielson, 111–38. Calgary: University of Calgary Press.

———. 1990. Mothering versus contract. In *Beyond self-interest*, ed. J. Mansbridge, 287–304. Chicago: University of Chicago Press.

Hertz, R. 1960. *Death and the right hand*. Trans. Rodney and Claudia Needham. Glencoe, Ill.: Free Press.

Hiatt, L. 1965. *Kinship and conflict*. Canberrra: Australian National University Press.

———. 1996. *Arguments about Aborigines: Australia and the evolution of social anthropology*. Cambridge, U.K.: Cambridge University Press.

Hibben, F. 1958. *Prehistoric man in Europe*. Norman: University of Oklahoma Press.

Hilbert, D. 1987. *Color and color perception*. Palo Alto, Calif.: Center for the Study of Language and Information, Leland Stanford University.

Hill, J. 1989. Concepts as units of cultural replication. *Journal of Social and Biological Structures* 12:343–55.

Hill, K., and M. Hurtado. 1996. *Aché life history: The ecology and demography of a foraging people*. New York: Aldine de Gruyter.

Hirota, T., M. Nara, M. Ohguri, and K. Hirota. 1992. Effect of diet and lifestyle on bone mass in Asian young women. *Journal of Clinical Nutrition* 55:1168–73.

Hobbes, T. 1946. *Leviathan (Parts 1 and 2)*. Indianapolis: Bobbs-Merrill.

Hoebel, E. A. 1949. Man in the primitive world: An introduction to anthropology. New York: McGraw-Hill.

Hollander, A. 1978. *Seeing through clothes: Fashioning ourselves, an intriguing new look at image-making*. New York: Avon.

Holt, E. 1958. *A documentary history of art*. Vol. 2. Garden City, N.Y.: Doubleday Anchor Books.

Holsinger, R., C. Jordan, and L. Levenson. 1971. *The creative encounter*. Glenview, Ill.: Scott, Foresman and Co.

Houston, M. 1920. *Ancient Egyptian, Mesopotamian, and Persian costumes and decoration*. London: Adams and Charles Black.

Howell, R. 1976. The population of the Dobe area !Kung. In *Karabari hunter-gatherers*, ed. R. Lee and I. Devore. Cambridge, Mass.: Harvard University Press.

Howells, W. 1948. *The heathens*. New York: Doubleday.

Hrdlicka, A. 1940. *Ear exototoses*. Washington, D.C.: Smithsonian Institution, Miscellaneous Collections.

Hrdy, S. 1977. *Langurs of Abu*. Cambridge, Mass.: Harvard University Press.

———. 1979a. The evolution of human sexuality: The latest work and the last. Review of Donald Symons, *The Evolution of Human Sexuality*. *Quarterly Review of Biology* 54:309–14.

————. 1979b. Infanticide among animals: A review, classification, and examination of the implications for the reproductive strategies of females. *Ethology and Sociobiology* 1:13–40.

————. 1986. Empathy, polyandry, and the myth of the coy female. In *Feminist approaches to science,* ed. R. Bleier, 119–46. New York: Pergamon.

————. 1999. *Mother nature: A history of mothers, infants, and natural selection.* New York: Pantheon Press.

Hudson, A. 1972. *Padju Epat: The Ma'anyan of Indonesian Bornero.* New York: Holt, Rinehart and Winston.

Humphrey, N. 1976. The social function of intellect. In *Growing points in ethology,* ed. P. P. G. Bateson and R. Hinde, 303–17. Cambridge, U.K.: Cambridge University Press.

Im Thurn, E. 1883. *Among the Indians of Guiana.* New York: Dover.

Ingold, T., ed. 1994. *Companion encyclopedia of anthropology.* London: Routledge.

Janowitz, B., J. Lewis, A. Parnell, F. Henawi, M. Yournis, and G. Serour. 1981. Breast-feeding and child survival in Egypt. *Journal of Biosocial Science* 13:287–97.

Jarvis, J. 1981. Eusociality in a mammal: Cooperative breeding in naked mole-rat colonies. *Science* 212:571–73.

Jastrow, M. 1911. Aspects of religious belief and practice in Babylonia and Assyria. New York: G. P. Putnam's Sons.

Jebb, Sir Richard. 1902. *The Antigone of Sophocles: With a commentary,* ed. E. Shuckburgh. Cambridge, U.K.: Cambridge University Press.

Jenness, D. 1922. *The life of the Copper Eskimo. Report of the Canadian Arctic expedition.* Vol. 12. Ottawa: F. A. Ackland.

Johansson, S. 1987. Neglect, abuse, and avoidable death: Parental investment and the mortality of infants and children in the European tradition. In *Child abuse and neglect,* ed. R. Gelles and J. Lancaster, 57–93. New York: Aldine de Gruyter.

Jolly, A. 1985. *The evolution of primate behavior.* 2d ed. New York: Macmillan Publishing Co.

Jolly, A. 1966. Lemur social behavior and primate intelligence. *Science* 153:501–6.

————. 1999. Lucy's legacy: Sex and intelligence in human evolution. Cambridge, Mass.: Harvard University Press.

Joyce, J. 1928. *The portrait of the artist as a young man.* New York: Modern Library.

Joyce, R. 1975. *The aesthetic animal: Man, the art-created art creator.* Hicksville, N.Y.: Exposition Press.

Kaberry, P. 1939. *Aboriginal women: Sacred and profane.* London: Routledge.

Kant, E. 1995. *Foundations of the metaphysics of morals.* Upper Saddle River, N.J.: Prentice-Hall.

Kaplan, H., K. Hill, J. Lancaster, and A. Hurtado. 2000. A theory of human life history evolution: Diet, intelligence, and longevity. *Evolutionary Anthropology* 9 (4): 156–84.

Keesing, R., and A. Strathern. 1998. *Cultural anthropology.* Fort Worth: Harcourt Brace College Edition.

Kehoe, A. 1991. No possible, probable shadow of doubt. *Antiquity* 65:129–31.

Keightley, D. 2000. *The ancestral landscape: Time, space, and community in late Shang China (ca. 1200–1045 BC)*. Berkeley: Institute of East Asian Studies, Center for Chinese Studies, University of California.

Key, C., and L. Aiello. 1999. The evolution of social organization. In *The evolution of culture: An interdisciplinary view*, ed. R. Dunbar, C. Knight, and C. Power, 15–33. New Brunswick, N.J.: Rutgers University Press.

Kipling, R. 1913. *The conundrum of the workshops: Departmental ditties and ballads and barrack room ballads*. New York: Macmillan and Co.

Klein, L., and L. Ackerman. 1995. Introduction to *Women and power in native North America*, ed. L. Klein and L Ackerman, 3–16. Norman: University of Oklahoma Press.

Klima, B. 1957. Übersicht über die jüngsten paläolithischen Forschungen in Mähren. *Quartär* 9:85–130.

Kluckhohn, C., and D. Leighton. 1962. *The Navajo*. Garden City, N.Y.: Doubleday, Anchor Books.

Klumpp, D. 1987. Masai art and society: Age and sex, time and space, cash and cattle. Ph.D. diss., Columbia University, New York.

Knight, C. 1991. *Blood relations: Menstruation and the origins of culture*. New Haven: Yale University Press.

Knight, C., R. Dunbar, and C. Power. 1999. An evolutionary approach to human culture. In *The evolution of culture: An interdisciplinary view*, ed. R. Dunbar, C. Knight, and C. Power, 1–11. New Brunswick, N.J.: Rutgers University Press.

Knobler, N. 1967. *The visual dialogue: An introduction to the appreciation of art*. New York: Holt, Rinehart and Winston.

Kohler, J. 1925. *On the prehistory of marriage*. Chicago: University of Chicago Press.

Konner, M. 1977. Infancy among the Kalahari Desert San. In *Culture and Infancy*, ed. S. Tulkin and A. Rosenfeld, 287–328. New York: Academic Press.

Konner, M., and M. Shostak. 1987a. Timing and management of birth among the !Kung: Biocultural interaction in reproductive adaptation. *Cultural Anthropology* 2 (1): 11–29.

———. 1987b. Adolescent pregnancy and childbearing. In *Schoolage pregnancy and childbearing: Biosocial dimensions*, ed. J. Lancaster and B. Hamburg. New York: Aldine.

Konner, M., and C. Worthman. 1980. Nursing frequency, gonadal function, and birth spacing among !Kung hunter-gatherers. *Science* 207:788–91.

Kraak, D., C. Bakker, and B. Mundwiler. 1999. Sexual selection in sticklebacks in the field: Correlates of reproductive, mating, and paternal success. *Behavioral Ecology* 10 (6): 696–706.

Kris, E., and O. Kurz. 1979. *Legend, myth, and magic in the image of the artist: A historical experiment*. New Haven: Yale University Press.

Kruuk, H. 1972. *The spotted hyena*. Chicago: University of Chicago Press.

Kummer, H., L. Daston, G. Gigerenzer, and J. Silk. 1997. The social intelligence hypothesis. In *Human by nature: Between biology and the social sciences*, ed. P. Wiengart, S. Mitchell, P. Richerson, and S. Maasen, 157–79. Mahwah, N.J.: Lawrence Erlbaum Associates.

Lack, D. 1954. *The natural regulation of animal numbers.* Oxford: Oxford University Press.

Lalvani, S. 1996. *Photography, vision, and the production of modern bodies.* New York: State University of New York Press.

Lancaster, J. 1975. *Primate behavior and the emergence of human culture.* New York: Holt, Rinehart and Winston.

Lancaster, J., J. Altmann, A. Rossi, and L. Sherrod, eds. 1987. *Parenting across the life span: Biosocial dimensions.* New York: Aldine de Gruyter.

Lancaster, J., and B. King. 1992. An evolutionary perspective on menopause. In *In her prime: A new view of middle-aged women,* ed. V. Kerns and J. Brown, 7–15. Chicago: University of Illinois Press.

Lancaster, J., and C. Lancaster. 1987. The watershed: Change in parental investment and family formation strategies in the course of human evolution. In *Parenting across the life span: Biosocial dimensions,* ed. J. Lancaster, J. Altmann, A. Rossi, and L. Sherrod, 187–205. New York: Aldine de Gruyter.

Lancaster, J., and P. Whitten. 1990. Sharing in human evolution. In *Anthropology: Contemporary perspectives.* 6th ed. Ed. P. Whitten and D. Hunter, 60–66. Glenview, Ill.: Scott, Foresman/Little, Brown Higher Education.

Langer, S. 1953. *Feeling and form.* New York: Charles Scribner's Sons.

———. 1958. *Problems of art.* New York: Charles Scribner's Sons.

Lawson, J. 1967. Sequelae of obstructed labor. In *Obstetrics and gynecology in the tropics and developing countries,* ed. J. J. Lawson and D. Steward, 172–202. London: Edward Arnold.

Lee, R. 1979. *The !Kung San: Men, women, and work in a foraging society.* Cambridge, U.K.: Cambridge University Press.

Lee, R., and I. DeVore. 1976. *Kalahari hunter-gatherers.* Cambridge, Mass.: Harvard University Press.

Leigh, R. 1937. Dental morphology and pathology of pre-Spanish Peru. *American Journal of Physical Anthropology* 23 (4): 429–49.

Leon, G. 1983. *Treating eating disorders: Obesity, anorexia nervosa, and bulimia.* Brattleboro, Vt.: Lewis Pub. Co.

Lewis-Williams, J. 1982. The economic and social context of southern San rock art. *Current Anthropology* 23 (4): 429–49.

Lewontin, R. 1970. The units of selection. *Annual Review of Ecological Systems* 1: 1–18.

Leroi-Gourhan, A. 1943. *Evolution et techniques: l'Homme et la matière.* Paris: Albin Michel.

———. 1945. *Evolution et techniques: Milieu et technique.* Paris: Albin Michel.

———. 1968. *The art of prehistoric man in Western Europe.* London: Thames and Hudson.

Levi-Strauss, C. 1962. *Totemism.* Boston: Beacon Press.

Lincoln, D. 1983. Physiological mechanisms governing the transfer of milk from mother to young. In *Symbiosis in parent-offspring interactions,* ed. L. Rosenblum and H. Moltz, 77–122. New York: Plenum Press.

Linne, S. 1940. Dental decoration in Aboriginal man. *Ethnos* 5:2–18.

Linton, R., and P. Wingert. 1946. *Arts of the South Seas.* New York: Museum of Modern Art.

Lippman, W. 1929. *A preface to morals.* New York: Macmillan Co.

Low, B. 1979. Sexual selection and human ornamentation. In *Evolutionary biology and human social behavior,* ed. N. Chagnon and W. Irons. North Scituate, Mass.: Duxbury Press.

———. 1989. Cross-cultural patterns in the training of children: An evolutionary perspective. *Journal of Comparative Psychology* 103:311–19.

———. 1998. The evolution of human life histories. In *Handbook of evolutionary psychology,* ed. C. Crawford and D. Krebs, 131–62. Mahwah, N.J.: Lawrence Erlbaum Associates.

———. 2000. *Why sex matters: A Darwinian look at human behavior.* Princeton, N.J.: Princeton University Press.

Lowie, R. 1929. *Culture and ethnology.* New York: P. Smith.

———. 1934. Some moot questions of social organization. *American Anthropologist,* 26:321–30.

———. 1937. *A history of ethnological theory.* New York: Farrar and Rinehart.

———. 1940. *An introduction to cultural anthropology.* New York: Farrar and Rinehart.

Lucie-Smith, E. 1992. *Art and civilization.* Englewood Cliffs, N.J.: Prentice-Hall.

Lumsden, C., and E. Wilson. 1981. *Genes, mind, and culture.* Cambridge, Mass.: Harvard University Press.

Maas, J. 1984. *Victorian painters.* New York: Abbeville Press.

MacCurdy, G. 1923. Human skeletal remains from the highlands of Peru. *American Journal of Physical Anthropology* 6 (3): 217–329.

MacDonald, K. 1992. Warmth as a development construct: An evolutionary analysis. *Child Development* 63:753–73.

———, ed. 1988. *Sociobiological perspectives on human development.* Hamburg: Springer-Verlag.

Major, J. 1970. *The age of the renaissance and reformation: A short history.* Philadelphia: Lippincott.

Malamuth, N., and M. Heilmann. 1998. Evolutionary psychology and sexual aggression. In *Handbook of evolutionary psychology,* ed. C. Crawford and D. Krebs, 515–42. Mahwah, N.J.: Lawrence Erlbaum Associates.

Malinowski, B. 1932. *The sexual life of savages.* 3d ed. London: Routledge.

———. 1955. Magic, science and religion. In *Magic, science and religion and other essays.* Garden City, N.Y.: Doubleday

Mallery. G. 1893. *Picture writing of the American Indians. Tenth annual report of the Bureau of American Ethnology to the secretary of the Smithsonian Institution, 1888–89.* Washington, D.C.: Bureau of American Ethnology.

Manser, J. 1946. Prehistoric dental inlays in Ecuador. *El Palacio* 5:111–15.

Marangou, C. 1992. *Eid-olia: Figurines et miniatures du Néolithique récent et du Bronze ancien en Grèce.* Oxford, U.K.: Tempus Reparatum.

Marquardt, C. 1984. *The tattooing of both sexes in Samoa.* Papakura, New Zealand: R. McMillan.

Marshack, A. 1981. On Paleolithic ochre and the early uses of color and symbol. *Current Anthropology* 22 (2): 188–91.

———. 1996. A middle Paleolithic symbolic composition from the Golan Heights: The earliest known depictive image. *Current anthropology* 37:357–65.

———. 1997. Paleolithic image making and symboling in Europe and the Middle East: A comparative review. In *Beyond art: Pleistocene art and symbol,* ed. M. Conkey, O. Soffer, D. Stratmann, and N. Jablonski, 53–92. San Francisco: California Academy of Sciences.

Maschner, H., and J. Patton. 1996. Kin selection and the origins of hereditary social inequality: A case study from the northern northwest coast. In *Darwinian Archaeologies,* ed. H. Maschner, 89–107. New York: Plenum Press.

Massola, A. 1971. *The Aborigines of south-eastern Australia: As they were.* Melbourne: William Heinemann Australia Pty. Ltd.

Matthews, W. 1977. *Ethnography and philology of the Hidatsa Indians.* Washington, D.C.: U.S. Geological and Geographical Survey, Miscellaneous Collections 7.

Mayers, W. 1974. *The Chinese readers manual: A handbook of biographical, historical, mythological, and general literary reference.* Shanghai: American Presbyterian Mission Press.

Maynard-Smith, J. 1978. *The evolution of sex.* Cambridge, U.K.: Cambridge University Press.

McAnany, P. 1995. *Living with the ancestors: Kinship and kingship in ancient Maya society.* Austin: University of Texas Press.

McCarthy 1939. "Trade" in Aboriginal Australia, and "trade" relations with Torres Strait, New Guinea, and Malaya. *Oceania* 9 (4): 405–38; 10 (2): 81–104; 10 (3): 171–95.

McEvilley, T. 1992. *Art and otherness: Crisis in cultural identity.* Kingston, N.Y.: McPherson and Co.

McFalls, J., and M. McFalls. 1984. *Disease and fertility.* Orlando, Fla.: Academic Press.

McGrew, W. 1977. Socialization and object manipulation of wild chimpanzees. In *Primate bio-social development,* ed. S. Chevalier-Skolnikoff and F. Poirier, 261–88. New York: Garland.

———. 1992. *Chimpanzee material culture.* Cambridge, U.K.: Cambridge University Press.

McKenna, J., and N. Bernshaw. 1995. Breastfeeding and infant-parent co-sleeping as adaptive strategies: Are they protective against SIDS? In *Breastfeeding: Biocultural perspectives,* ed. P. Stuart-Macadem and K. Dettwyler, 265–303. New York: Aldine de Gruyter.

McNeill, R., and G. Newton. 1965. Cranial base morphology in association with intentional cranial vault deformation. *American Journal of Physical Anthropology* 23:241–53.

Mead, S., and B. Kernot, eds. 1983. *Art and artists of Oceania.* Palmerston North, New Zealand: Dunmore Press.

Meaney, Michael. 2001. Maternal care may leave brain legacy. *Science News,* November 3, 2001.

Meggitt, M. 1962. *Desert people: A study of the Walbiri Aborigines of central Australia.* Sydney: Angus and Robertson.

———. 1965a. The association between Australian Aborigines and Dingoes. In *Man, culture, and animals: The role of animals in human ecological adjustments,* ed. A. Leeds and A. Vayda, 7–26. Washington, D.C.: American Association for the Advance of Science.

———. 1965b. *The lineage system of the Mae Enga of New Guinea.* Edinburgh: Oliver and Boyd.

———. 1965c. The Mae Enga of the western Highlands. In *Gods, ghosts, and men in Melanesia,* ed. P. Lawrence and M. Meggitt, 105–31. New York: Oxford.

———. 1965d. *Desert people.* Chicago: University of Chicago Press.

Mellars, P. 1989. Major issues in the emergence of modern humans. *Current Anthropology* 30:349–85.

———. 1990. *The emergence of modern humans: An archaeological perspective.* Edinburgh: Edinburgh University Press.

Mellars, P., and C. Stringer, eds. 1969. *The human revolution.* Edinburgh: University of Edinburgh Press.

Middleton, J. 1960. *Lugbara religion.* London: Oxford University Press, International African Institute.

Mildvan, A., and B. Strehler. 1960. A critique of theories of mortality. In *The biology of aging,* ed. B. Strehler, 45–78. New York: American Institute of Biological Sciences.

Millar, J. 1979. Energetics of lactation in *Peromyscus maniculatus. Canadian Journal of Zoology* 57:1016–19.

Miller, G. 1999. Sexual selection for cultural displays. In *The evolution of culture: An interdisciplinary view,* ed. R. Dunbar, C. Knight, and C. Power, 71–91. New Brunswick, N.J.: Rutgers University Press.

———. 2000. *The mating mind: How sexual choice shaped the evolution of human nature.* New York: Doubleday.

Mitchell, W. 1998. Women's hierarchies of age and suffering in an Andean community. In *Women among women: Anthropological perspectives on female age hierarchies,* ed. J. Dickerson-Putman and J. Brown, 52–64. Urbana: University of Illinois.

Mithen, S. 1998. *The prehistory of the mind.* London: Thames and Hudson.

Montagu, A. 1957. *The reproductive development of the female.* New York: Julian Press.

Moore, A. C. 1989. The land as religious symbol: A view from New Zealand. In *Doing theology with people's symbols and images,* ed. Yeow Choo Lak and J. C. England, 84–100. ATESEA Occasional Papers No. 8, Association for Theological Education in S. E. Asia, Singapore.

———. 1995. *Arts in the religions of the Pacific: Symbols of life.* London: Pinter Publishers.

Morphy, H. 1984. *Journey to the crocodiles nest.* Canberra: Australian Institute of Aboriginal Studies.

———. 1991. *Ancestral connections: Art and an Aboriginal system of knowledge.* Chicago: University of Chicago Press.

————. 1992. From dull to brilliant: The aesthetics of spiritual power. In *Anthropology, art, and aesthetics,* ed. J. Coote and A. Shelton, 181–208. Oxford: Clarendon.

————. 1993. On representing ancestral beings. In *Arts of Africa, Oceania, and the Americas: Selected readings,* ed. J. Berlo and L. Wilson, 245–58. Englewood Cliffs, N.J.: Prentice-Hall

Morris, D. 1962. *The biology of art.* London: Methuen.

Moseley, M. 1992. *The Incas and their ancestors: The archaeology of Peru.* London: Thames and Hudson.

————. 1995. The Incas and their ancestors: Archaeology of Peru. *Antiquity* 59 (263): 398–402.

Mountford, C. 1958. *The Tiwi: Their art, myth, and ceremony.* London: Phoenix House.

Mulvaney, J., and J. Kamminga. 1999. *Prehistory of Australia.* Washington: Smithsonian Institution Press.

Mundinger, P. 1980. Animal cultures and a general theory of cultural evolution. *Ethology and Sociobiology* 1:183–223.

Munn, N. 1973. *Walbiri iconography.* Ithaca, N.Y.: Cornell University Press.

Munro, T. 1949. *The arts and their interrelationships.* New York: Liberal Arts Press.

Murdock, G. 1971. Anthropology's mythology: The Huxley memorial lecture. *Journal of the Royal Anthropological Institute* 17:23.

Myers, D. 1993. *Social psychology.* 4th ed. New York: McGraw-Hill.

National Gallery of Art. 2000. *Modern art and America: Alfred Stieglitz and his New York Galleries.* www.nga.gov/xio/maa_stiegpr.htm.

Nauert, C. 1977. *The age of renaissance and reformation.* Hinsdale, Ill.: Dryden Press.

Nelson, S. 1997. *Gender in archaeology.* Walnut Creek, Calif.: AltaMira Press.

————. 2001. Diversity of the Upper Paleolithic "Venus" figurines and archaeological mythology. In *Gender in cross-cultural perspective,* ed. C. Brettell and C. Sargent, 82–89. Upper Saddle River, N.J.: Prentice-Hall.

Nelson, M., and H. Evans. 1961. Dietary requirements for lactation in the rat and other laboratory mammals. In *Milk: The mammary gland and its secretion,* vol. 2., ed. S. Kon and A. Cowie. New York: Academic Press.

Nesse, R., and G. Williams. 1994. *Why we get sick.* New York: Times Books.

Newberger, E. 1999. *The men they will become: The nature and nurture of the male character.* Cambridge, Mass.: Perseus Books.

O'Brien, E. 1984. What was the Acheulean hand axe? *Natural History* 93 (7): 20–23.

Ohnuki-Tierney, E. 1974. *The Ainu of the northwest coast of southern Sakhalin.* New York: Holt, Rinehart and Winston.

Orquera, L. 1984. Specialization and the Middle/Upper Paleolithic transition. *Current Anthropology* 25:73–98.

Ortiz, G. 1966. *In pursuit of the absolute: Art of the ancient world.* Rev. ed. Bern, Switzerland: author.

Otte, P. 1974. Effects and functions in the evolution of signaling systems. *Annual Review of Ecology and Systematics* 5:1259–68.

Palmer, C., B. Fredrickson, and C. Tilley. 1997. Categories and gatherings: Group selection and the mythology of cultural anthropology. *Evolution and Human Behavior* 18 (5): 291–308.

Palmer, C., and L. Steadman. 1997. Human kinship as a descendant-leaving strategy: A solution to an evolutionary puzzle. *Journal of Social and Evolutionary Systems* 20:39–52.

Papathanassopoulos, G., ed. 1996. *Neolithic culture in Greece.* Athens: Nicholas P. Goulandris Foundation, Museum of Cycladic Art.

Parezo, N. 1983. *Navajo sandpainting: From religious act to commercial art.* Tucson: University of Arizona Press.

Parkinson, S. 1784. *A journal of a voyage to the South Seas in H.M.* Ship Endeavour. London: Charles Dilly and James Philip.

Parsons, E. 1919. *Notes on Zuñi.* Lancaster, Penn.: American Anthropological Association.

Patton, S., and R. Jensen. 1976. *Biomedical aspects of lactation.* Oxford: Pergamon Press.

Pavelka, M. 1998. The nonhuman primate perspective: Old age, kinship, and social partners in a monkey society. In *Women among women: Anthropological perspectives on female age hierarchies,* ed. J. Dickerson-Putman and J. Brown, 89–99. Urbana: University of Illinois Press.

Pawlowski, B., C. Lowen, and R. Dunbar. 1998. Neocortex size, social skills, and mating success in primates. *Behavior* 135:357–68.

Pernet, H. 1992. *Ritual masks: Deceptions and revelations.* Columbia: University of South Carolina Press.

Piddington, R. 1950. *An introduction to social anthropology.* Vol. 1. Edinburgh: Oliver and Boyd.

Pinker, S. 1997. *How the mind works.* New York: W. W. Norton and Co.

Pisacane, A., G. Grillo, M. Cafiero, C. Simeone, A. Coppola, B. Scarpellino, and G. Mazzarella. 1992. Role of breastfeeding in paralytic poliomyelitis. *British Medical Journal* 305:1367.

Plato. 1977. Art is imitation. From *The republic.* Trans. Benjamin Jowett. In *Aesthetics: A critical anthropology,* ed. G. Dickie and R. Sciafani. New York: Bobbs Merrill.

Podelsky, A., and P. Brown. 1997. *Applying cultural anthropology.* 2d ed. Mountain View, Calif.: Mayfield.

Polhemus, T., ed. 1978. *The body reader: Social aspects of the body.* New York: Pantheon Books.

Popham, A. 1945. *The drawings of Leonardo da Vinci.* New York: Harcourt, Brace and World.

Praslov, N. 1985. L'art du Paléolithique Supérieur à l'est de l'Europe. *L'Anthropologie* 89:181–92.

Pridelle, H., W. Lenz, D. Young, and C. Stevenson. 1952. Poliomyelitis in pregnancy and the puerperium. *American Journal of Obstetrics and Gynecology* 63:408–13.

Pulliam, H., and C. Dunford. 1980. *Programmed to learn.* New York: Columbia University.

Quilter, J. 1989. *Life and death at a Paloma society and mortuary practices in a preceramic a Peruvian village.* Iowa City: University of Iowa Press.

Radcliffe-Brown, A. 1922. *The Andaman Islanders.* Cambridge, U.K.: Cambridge University Press.

———. 1952. *Structure and function in primitive society.* Glencoe, Ill.: Free Press.

Radin, P. 1953. *The world of primitive man.* New York: H. Schuman.

———. 1963. *Social anthropology.* New York: McGraw-Hill.

Ravenholt, R., and J. Chao. 1974. *World fertility trends. Population reports.* Washington, D.C.: George Washington University Department of Medical and Public Affairs.

Reay, M. 1959. *The Kuma: Freedom and conformity in the New Guinea Highlands.* Melbourne: Melbourne University Press.

Reichard, G. 1950. *Navajo religion: A study of symbolism.* Princeton, N.J.: Princeton University Press.

Reichel-Dolmatoff, G. 1961. Anthropomorphic figurines from Colombia: Their magic and art. In *Essays in Pre-Colombian art and archaeology,* ed. S. Lathrop et al., 229–41. Cambridge, Mass.: Harvard University Press.

Reynolds, V. 1994. Kinship in nonhuman and human primates. In *Hominid culture in primate perspective,* ed. D. Quiatt and J. Itani., 137–65. Niwot, Colo.: University Press of Colorado.

Rice, P. 1981. Prehistoric Venuses: Symbols of motherhood or womanhood? *Journal of Anthropological Research* 37 (4): 402–14.

Richards, A. 1956. *Chisungu: A girl's initiation ceremony among the Bemba of northern Rhodesia.* London: Faber and Faber.

Riding, A. 1985. *Distant neighbors: A portrait of the Mexican.* New York: Alfred A. Knopf.

Ridley, M. 1996. *The origins of virtue.* London: Viking.

———. 2001. Re-reading Darwin. *Prospect Magazine.* Selected Features, Arts and Books (August/September): www.prospect_magazine.co.uk.

Roche, A., ed. 1979. *Secular trends in human growth, maturation, and development: Chicago.* Chicago: University of Chicago Press for the Society for Research in Child Development.

Rogers, A., and J. Berntsen. 1992. Evolution of menopause. Presentation given at the 1992 Conference on Human Behavior and Evolution, University of New Mexico, Albuquerque, July 22–25.

Rogers, S. 1975. Artificial deformation of the head: New World examples of ethnic mutilation and notes on its consequences. San Diego, Calif.: Museum of Man.

Romero, J. 1970. Dental mutilations, trephination, and cranial deformation. In *Handbook of Middle American Indians,* ed. J. Wauchope. Austin: University of Texas Press.

Romero Molina, J. 1986. Nuevos datos sobre la mutilación dentaria en Mesoamérica. *Anales de Anthropología* 23:239.

Ronnekleiv, O., S. Ojeda, and S. McCann. 1979. Undernutrition, puberty, and the development of estrogen positive feedback in the female rat. *Biology and Reproduction* 19:424.

Roosevelt, A. 1988. Interpreting certain female images in prehistoric art. In *The role of gender in Precolumbian art and architecture,* ed. V. Miller, 1–34. Landham, Md.: University Press of America.

Rosenblum, L., and H. Moltz, eds. 1983. *Symbiosis in parent-offspring interactions.* New York: Plenum Press.

Ross, C., and K. Jones. 1999. Socioecology and the evolution of primate reproductive strategies. In *Comparative primate socioecology,* ed. P. Lee, 73–110. Cambridge, U.K.: Cambridge University Press.

Rudgley, R. 1999. *The lost civilizations of the Stone Age.* New York: Free Press.

Rueschenberg, E., and R. Buriel. 1995. Mexican American family functioning and acculturation: A family systems perspective. In *Hispanic psychology: Critical issues in theory and research,* ed. A. Padilla, 18–25. Thousand Oaks, Calif.: Sage Publications.

Russell, P. 1991. Men only? The myths about European Paleolithic artists. In *The archaeology of gender: Proceedings of the 22nd annual conference of the archaeological association of the University of Calgary,* ed. D. Walde and N. Willows, 346–51. Calgary: University of Calgary.

Ruyle, E. 1973. Genetic and cultural pools: Some suggestions for a unified theory of biocultural evolution. *Human Ecology* 1:201–15.

Saggs, H. 1962. *The greatness that was Babylon: A sketch of the ancient civilizations of the Tigris-Euphrates Valley.* New York: New American Library.

———. 1984. *The might that was Assyria.* London: Sidgwick and Jackson.

Sanchez-Tranquilino, M. 1985. Mexican American graffiti and Chicano murals in east Los Angeles, 1972–1978. In *Looking high and low: Art and cultural identity,* ed. B. Bright and L. Bakewell, 55–88. Tucson: University of Arizona Press.

Sanders, N. 1968. *Prehistoric art in Europe.* Baltimore: Penguin.

Santos, F., and F. Barcley. 1984. *Guía etnográfica de la Alta Amazonía.* Vol. 1. Quito: FLASCO Sede Ecuador.

Santos Granero, F. 1991. *The power of love: The moral use of knowledge amongst the Amuesha of central Peru.* London School of Economics Monographs on Social Anthropology, 62. London: Athlone Press.

Saxe, A. 1971. Social dimensions of mortuary practices in a Mesolithic population from Wadi-Halfa, Sudan. In *Approaches to the social dimensions of mortuary practices,* ed. J. Brown, 39–57. Issued as *American Antiquity* 35 (3). Memoirs of the Society for American Archaeology, No. 25.

Sayce, A. 1900. *Babylonians and Assyrians: Life and customs.* New York: Charles Scribner's Sons.

Schaller, G. 1972. *The Serengetti lion.* Chicago: University of Chicago Press.

Scheper-Hughes, N. 1991. Lifeboat ethics: Mother love and child death in Brazil. In *Applying cultural anthropology,* ed. A. Podelefsky and P. Brown. Mountain View, Calif.: Mayfield.

Schneebaum, T. 1985. *Asmat images: From the collection of the Asmat Museum of Culture and Progress: Gambaran Asmat.* Minneapolis: Koleski Museum Kebudayaan dan Kamajuan.

Schuster, C., and E. Carpenter. 1996. *Patterns that connect: Social symbolism in ancient and tribal art.* New York: Harry N. Abrams.

Schwartz, M., D. Boness, C. Schaeff, P. Majluf, E. Perry, and R. Fleischer. 1999. Female-solicited extrapair matings in Humboldt penguins fail to produce extrapair fertilizations. *Behavioral Ecology* 10 (3): 242–50.

Schweitzer, M. 1999. *American Indian grandmothers.* Albuquerque: University of New Mexico Press.

Seeger, A. 1973. The meaning of body ornaments: A Suya example. *Ethnology* 14 (3): 211–24.

Seligman, G. 1952. *Oh fickle taste, or objectivity in art.* Toronto: Bond Wheelwright Co.

Shannon, C. 1989. *The politics of the family.* New York: Peter Long.

Shapiro, W. 1981. *Miwuyt marriage.* Philadelphia: Institute for the Study of Human Issues.

Shaw, A. 1974. *A Pima past.* Tucson: University of Arizona Press.

Sherbondy, J. 1982. El ragadio, los lagos y los mítos de orígen. *Allpanchis* 20:3–32.

Simkiss, K. 1967. *Calcium in reproductive physiology.* London: Chapman and Hall.

Simmons, D. 1986. *Ta moko: The art of Maori tattoo.* Auckland: Reed Methuen.

Smidt, D. 1993. *Asmat art: Woodcarvings of southwest New Guinea.* Leiden: Rijksmuseum voor Volkenkunde.

Smith, T., and T. Polacheck. 1981. Reexamination of the life table for northern fur seals with implications about population regulatory mechanisms. In *Dynamics of large mammalian populations,* ed. C. Fowler and T. Smith, 99–120. New York: Wiley.

Soffer, O. 1987. *The Pleistocene Old World: Regional perspectives.* New York: Plenum.

————. 1993. Upper Paleolithic adaptations in central an eastern Europe and Man-Mammoth interactions. In *From Kostenki to Clove: Upper Palotlithic-Paleo-Indian adaptations,* ed. O. Soffer and N. Praslov, 51–65. New York: Plenium.

Soffer, O., J. Adovasio, and D. Hyland. 2000. The Venus figurines: Textiles, basketry, gender, and status in the Upper Paleolithic. *Current Anthropology* 14:511–37.

Soffer, O., and M. Conkey. 1997. Studying ancient visual cultures. In *Beyond art: Pleistocene art and symbol,* ed. M. Conkey, O. Soffer, D. Stratmann, and N. Jablonski, 1–16. San Francisco: California Academy of Sciences.

Spencer, B., and F. Gillen. 1912. *Across Australia.* London: Macmillan and Co., Ltd.

————. 1938. *Native tribes of central Australia.* London: Macmillan.

————. 1966. *The Arunta: A study of a Stone Age people.* Vol. 1. Oosterhout, The Netherlands: Anthropological Publications.

————. 1969. *Northern tribes of central Australia.* Oosterhout, The Netherlands: Anthropological Publications.

Spicer, E. H. 1971. Persistent cultural systems: A comparative study of identity systems that can adapt to contrasting environments. *Science* 174:795–800.

————. 1980. *The Yaquis: A cultural history.* Tucson: University of Arizona.

Spielmann, K. 1989. A review: Dietary restrictions on hunter-gatherer women and the implications for fertility and infant mortality. *Human Ecology* 17:321–45.

Spindler, K., H. Wilfing, E. Rastbichler-Zissernig, D. zur Nedden, and H. Nothdurfter, eds. 1996. *The man in the ice.* Vol. 3, *Human mummies: A global survey of their status and the techniques of conservation.* New York: Springer Wien.

Statler, R. 1995. Women's status among the Muskogee and Cherokee. In *Women and power in North America,* ed. L. Klein and L. Ackerman, 214–29. Norman: University of Oklahoma Press.

Steadman, L. 1994. *Social behavior and sacrifice.* Paper presented at the Human Behavior and Evolution Society Meetings, June 16, Ann Arbor, Mich.

———.1995. *Traditions are not explained by r.* Paper presented at the Human Behavior and Evolution Society Meetings, Santa Barbara, Calif., July 2.

———. 1997. *Kinship hierarchy: The basis of cooperation?* Paper presented at the Human Behavior and Evolution Society Meetings, Tucson, Ariz., June 6.

Steadman, L., and C. Merbs. 1982. Kuru and cannibalism? *American Anthropologist* 84:611–27.

Steadman, L., and C. Palmer. 1995. Religion as an identifiable traditional behavior subject to natural selection. *Journal of Social and Evolutionary Systems* 18:149–64.

Steadman, L., C. Palmer, and C. Tilly. 1996. The universality of ancestor worship. *Ethnology* 35:63–76.

Stott, D. 1973. Follow-up study from birth of the effects of prenatal stresses. *Developmental Medicine and Child Neurology* 15:770–87.

Strassman, B. 1981. Sexual selection, parental care, and concealed ovulation in humans. *Ethology and Sociobiology* 2:31–40.

Strathern, A., and M. Strathern. 1971. *Self-decoration in Mount Hagen.* Toronto: University of Toronto Press.

Strommenger, E. 1962. *Fünf jahrtausende Mesopotamien* (Five thousand years of the art of Mesopotamia). München: Hirmer.

Stuart, D. 1988. Blood symbolism in Maya iconography. In *Maya incongraphy,* ed. E. Benson and G. Griffin, 175–221. Princeton: Princeton University Press.

Stuart, J., and E. Rawski. 2001. *Worshiping the ancestors: Chinese commemorative portraits.* Published by the Freer Gallery of Art and the Arthur M. Sacker Gallery, Smithsonian Institution, Washington, D.C., in association with Stanford University.

Stuart-Fox, M. 1986. The unit of replication in socio-cultural evolution. *Journal of Social and Biological Structures* 9:67–97.

Stubbs, D. 1974. *Prehistoric art of Australia.* New York: Charles Scribner's Sons.

Super, C. 1981. Behavioral development in infancy. In *Handbook of cross-cultural human development,* ed. R. Munroe and B. Whiting, 181–90. New York: Garland STPM Press.

Surbey, M. 1990. Family composition, stress, and human menarche. In *The socioendocrinology of primate reproduction,* ed. T. Ziegler and F. Bercovitch. West Sussex, U.K.: Wiley-Liss.

Symons, D. 1979. *The evolution of human sexuality.* New York: Oxford University.

Tacitus. 1942. *Complete works.* New York: Modern Library.

Talalay, L. 1993. *Deities, dolls, and devices: Neolithic figurines from Franchthi cave.* Bloomington: Indiana University Press.

Tanner, J. 1962. *Growth at adolescence.* Oxford: Blackwell.

Tavis, C., and S. Sadd. 1977. *The Redbook report on female sexuality.* New York: Delacorte Press.

Taylor, P., and L. Aragon. 1991. *Beyond the Java Sea: Art of Indonesia's outer islands.* Washington, D.C.: National Museum of Natural History, Smithsonian Institution.

Thévoz, M. 1984. *The painted body.* New York: Skira Rizzoli.

Thompson, R. 1973. Yoruba artistic criticism. In *The traditional artist in African societies,* ed. W. d'Acevedo, 19–61. Bloomington: Indiana University Press.

———. 1993. An aesthetic of the cool. In *Arts of Africa, Oceania, and the Americas,* ed. J. Berlo and L. Wilson, 22–47. Englewood Cliffs, N.J.: Prentice Hall.

Thomsen, M. 1996. *My two wars.* South Royalton, Vt.: Steerforth Press.

Thornhill, R. 1998. Darwinian aesthetics. In *Handbook of evolutionary psychology,* ed. C. Crawford and D. Krebs, 543–72. Mahwah N.J.: Lawrence Erlbaum Associates.

———. n.d. Darwinian aesthetics informs traditional aesthetics. Unpublished manuscript.

Thornhill, R., and C. Palmer. 2000. *A natural history of rape.* Cambridge, Mass.: MIT Press.

Tice, K. 1995. *Kuna crafts, gender, and the global economy.* Austin: University of Texas Press.

Tolstoy, L. 1977. Art as the communication of feeling. Trans. Aylmer Maude. In *Aesthetics: A critical anthropology,* ed. G. Dickie and R. Sciafani. New York: Bobbs Merrill.

Tomásková, S. 1997. Places of art: Art and archaeology in context. In *Beyond art: Pleistocene art and symbol,* M. Conkey, O. Soffer, D. Stratmann, and N. Jablonski, 265–88. San Francisco: California Academy of Sciences.

Tomassoni, I. 1968. *Pollock: The life and works of the artist.* New York: Grosset and Dunlap.

Tonkinson, Robert. 1978. *The Mardujara Aborigines: Living the dream in Australia's desert.* York: Holt, Rinehart and Winston.

Tooby, J., and L. Cosmides. 1989. Evolutionary psychology and the generation of culture. Pt. 1, Theoretical considerations. *Ethology and Sociobiology* 10:29–50.

———. 1990. On the universality of human nature and the uniqueness of the individual: The role of genetics and adaptation. *Journal of Personality* 58:17–67.

———. 1992. The psychological foundations of culture. In *The adapted mind,* ed. J. Barkow, L. Cosmides, J. Tooby. New York: Oxford University Press.

Tozzer, A. 1941. *Landa's relación de las cosas de Yucatán: A translation.* Papers of the Peabody Museum of Archaeology and Ethnology, 18. Cambridge, Mass.: Harvard University Press.

Trevathan, W. 1987. *Human birth: An evolutionary perspective.* New York: Aldine.

Trilling, L. 1971. Art and neurosis. In *The creative encounter,* ed. R. Holsinger, C. Jordan, C. Levenson, and L. Levenson 1971, 94–108. Glenview, Ill.: Scott, Foresman and Co.

Trivers, R. 1971. The evolution of reciprocal altruism. *Quarterly Review of Biology* 46:35–57.

———. 1972. Parental investment and sexual selection. In *Sexual selection and the descent of man,* ed. B. Campbell, 136–79. Chicago: University of Chicago Press.

———. 1974. Parent-offspring conflict. *American Zoologist* 14:249–64.

———. 1999. As they would do to you. *Skeptic* 6:81–83.

Tuchman, B. 1978. *A distant mirror: The calamitous fourteenth century.* New York: Ballantine Books.

Turke, P. 1988. Helpers at the nest: Childcare networks in Ifaluk. In *Human Reproductive Behavior: A Darwinian Perspective,* ed. L. Betzig, M. Borgerhoff Mulder, and P. Turke, 173–88. *Ethology and Sociobiology* 6:79–87.

———. 1989. Evolution and the demand for children. *Population and Development Review* 15:61–87.

Turner, T. 1969. Tchikrin: A central Brazilian tribe and its symbolic language of body ornament. *Natural History* 78 (8): 50–70.

Turner, F. 1991. *Beauty: The value of values.* Charlottesville: University Press of Virginia.

Turner, V. 1969. *The ritual process.* London: Routledge and Kegan Paul.

Tylor, E. 1960. *Anthropology.* Ann Arbor: University of Michigan Press.

Ucko, P. 1977. *Form in indigenous art: Schematisation in the art of aboriginal Australia and prehistoric Europe.* Prehistory and Material Culture Series, no. 13. Canberra: Australian Institute of Aboriginal Studies.

Udall, J., M. Dixon, and A. Newman. 1985. Liver disease in alpha-1-antitrysin deficiency: A retrospective analysis of early breast vs. bottle feeding. *Journal of the American Medical Association* 253:2679–82.

United Nations. 1985. *Socio-economic differential in child mortality in developing countries.* New York: United Nations.

Van Baal, J. 1981. *Man's quest for partnership.* The Netherlands: Van Gorcium.

Van Gennep, A. 1960. *The rites of passage.* Chicago: Phoenix Books, University of Chicago Press.

Van Lawick Goodall, J. and H. 1970. *Innocent killers.* London: Collins.

Van Zon, A., and W. Eling. 1980. Depressed malarial immunity during pregnancy. *American Journal of Obstetrics and Gynecology* 47:495–505.

Vasari, G. 1986. *The great masters: Giotto, Botticelli, Leonardo, Raphael, Michelangelo, Titian.* Trans. G. Du C. de Vere. New York: Hugh Lauter Levin Associates.

Vaughn, T. 1978. *Mammalogy.* Philadelphia: Saunders College Publishing.

Vining, D. 1982. Fertility differentials and the status of nations. *Mankind Quarterly* 22:311–53.

Vinnicomb, P. 1976. *People of the Eland: Rock paintings of the Drakensberg Bushmen as a reflection of their life and thought.* Pietermaritzburg: University of Natal Press.

Vogel, S. 1993. The Buli master and other hands. In *Arts of Africa, Oceania, and the Americas: Selected readings,* ed. J. Berlo and L. Wilson, 68–75. Englewood Cliffs, N.J.: Prentice-Hall.

Vogt, E. 1964. *Ancient Maya concepts in Contemporarya Zinacantan religion. VI congrès international des sciences anthropologiques et ethnologiques II (2).* Paris: Musée de l'Homme.

Wadley, R. 1999. Disrespecting the dead and the living: Iban ancestor worship and the violation of mourning taboos. *Journal of the Royal Anthropological Institute* 5:595–610.

Wagner, R. 1991. The use of extrapair copulations for mate appraisal by razorbills, *Alca torda. Behavioral Ecology* 2:198–203.

Walens, S. 1993. The weight of my name is a mountain of blankets: Potlatch ceremonies. In *Arts of Africa, Oceania, and the Americas: Selected readings,* ed. J. Berlo and L. Wilson, 184–200. Englewood Cliffs, N.J.: Prentice-Hall.

Walker, D. 1967. *Mutual cross-utilization of economic resources in the Plateau.* Pullman: Washington State University, Laboratory of Anthropology, Report of Investigations No. 41.

*Washington Post.* 1999. Findings: Depression affects newborns. May 19, A106.

Watchman, A. 1997. Paleolithic marks: Archaeometric perspectives. In *Beyond art: Pleistocene art and symbol,* ed. M. Conkey, O. Soffer, D. Stratmann, and N. Jablonski, 19–36. San Francisco: California Academy of Sciences.

Watts, E. 1985. Adolescent growth and development of monkeys, apes, and humans. In *Nonhuman primate models for growth and development,* ed. E. S. Watts. New York: Alan R. Liss.

———. 1999. The origin of symbolic culture. In *The evolution of culture: An interdisciplinary view,* ed. R. Dunbar, C. Knight, and C. Power, 34–49. New Brunswick, N.J.: Rutgers University Press.

Weatherhead, P. 1994. Mixed mating strategies by females may strengthen the sexy son hypothesis. *American Naturalist* 113:201–8.

Weingart, P., R. Boyd, W. Durham, and P. Richerson. 1997. Units of culture: Types of transmission. In *Human by nature: Between biology and the social sciences.* P. Wiengart, S. Mitchell, P. Richerson, and S. Maasen, 299–325. Mahwah, N.J.: Lawrence Erlbaum Associates.

Weissner, P. 1977. *Hxaro: A regional system of reciprocity for reducing risk among the !Kung San.* Ph.D. diss., University of Michigan.

———. 1983. Style and social information in Kalahari San projectile points. *American Antiquity* 48 (2): 263–75.

———. 1984. Reconsidering the behavioral basis for style: A case study among the Karahari San. *Journal of Anthropological Archaeology* 3:190–234.

West, J. 1991. Grutas in Spanish southwest. In *Hecho en Tejas: Texas-Mexican folk arts and crafts,* ed. J. Graham, 263–77. Denton: University of North Texas Press.

West, M., and M. Konner. 1976. The role of father: An anthropological perspective. In *The role of the father in child development,* ed. M. Lamb, 185–217. New York: John Wiley and Sons.

Westheim, P. 1950. *Arte antiguo de México.* Mexico, D.F.: Fondo de Cultura Economica.

Weltfish, G. 1953. *The origins of art.* Indianapolis: Bobbs-Merrill.

Westneat, D., P. Sherman, and M. Morton. 1990. The ecology and evolution of extra-pair copulations in birds. *Current Ornithology* 7:331–69.

Whallon, R. 1989. Elements of culture change in the Later Paleolithic. In *The human revolution,* ed. P. Mellars and C. Stringer, 433–54. Edinburgh: Edinburgh University Press.

White, D. 1993. Kwakiutl transformation masks. In *Arts of Africa, Oceania, and the*

*Americas: Selected readings,* ed. J. Berlo and L. Wilson, 266–95. Englewood Cliffs, N.J.: Prentice-Hall.

White, I. 1970. Aboriginal women's status: A paradox resolved. In *Women's role in Aboriginal society,* ed. F. Gale, 21–29. Canberra: Australian Institute of Aboriginal Studies.

White, M., and J. Gribbin. 1995. *Darwin: A life in science.* New York: Dutton.

White, R. 1992. Rethinking the Middle/Upper Paleolithic transition. *Current Anthropology* 23:169–92.

———. 1997. Substantial acts: From materials to meaning in Upper Paleolithic representation. In *Beyond art: Pleistocene art and symbol,* ed. M. Conkey, O. Soffer, D. Stratmann, and N. Jablonski, 93–121. San Francisco: California Academy of Sciences.

Whiting, B., and C. Edwards. 1988. *Children of different worlds: The formation of social behavior.* Cambridge, Mass.: Harvard University Press.

Wild, S., ed. 1986. *Rom: An Aboriginal ritual of diplomacy.* Canberra: Australian Institute of Aboriginal Studies.

Willett, F. 1971. Ife in Nigerian art. In *Anthropology and art: Readings in cross-cultural aesthetics,* ed. C. Otten, 383–404. Austin: University of Texas Press.

Williams, G. 1957. Pleiotropy, natural selection, and the evolution of senescence. *Evolution* 11:32–39.

———. 1966. *Adaptation and natural selection:* Princeton, N.J.: Princeton University Press.

———. 1975. *Sex and evolution.* Princeton, N.J.: Princeton University Press.

———. 1992. *Natural selection: Domains, levels, and challenges.* New York: Oxford University Press.

Wilson, D., and E. Sober. 1994. Reintroducing group selection to the human behavioral sciences. *Behavioral and Brain Sciences* 17:585–654.

Wilson, E. O. 1971. *The insect societies.* Cambridge, Mass.: Harvard University Press.

———. 1975. *Sociobiology: The new synthesis:* New York: Harvard University Press.

———. 1998. *Consiliance: The unity of knowledge.* New York: Knopf.

Wilson, M. 1957. *Rituals of kinship among the Nyakyusa.* London: Oxford University Press.

Wojcik, D. 1995. *Punk and neo-tribal body art.* Jackson: University of Mississippi Press.

Wood, J. 1994. *Dynamics of human reproduction: Biology, biometry, demography.* New York: Aldine de Gruyter.

Wood, W. R. 1962. A stylistic and historical analysis of shoulder patterns on Plains Indian pottery. *American Antiquity* 28 (1): 25–40.

———. 1980. Plains trade in prehistoric and protohistoric intertribal relations. In *Anthropology on the Great Plains,* ed. W. R. Wood and M. Liberty, 98–109. Lincoln: University of Nebraska Press.

———. 1991. Ecology and Great Plains studies. *Review of Archaeology* 12 (2): 30–34.

Woolf, V. 1957. *A room of one's own.* New York: Harcourt Brace.

Wrangham, R. 1980. An ecological model of female-bonded primate groups. *Behavior* 75:262–300.

Wreschner, E. 1980. Red ochre and human evolution: A case for discussion. *Current Anthropology* 21:631–44.

Wright, J., L. Steadman, C. Palmer, and P. Stamile. 1997. More kin: An evolutionary explanation of marriage. Paper presented at the Human Behavior and Evolution Society, June, University of Arizona.

Wright, R. 1994. *The moral animal.* New York: Pantheon.

Wu, H. 1989. *The Wu Liang Shrine: The ideology of early Chinese pictorial art.* Stanford: Stanford University Press.

Wyman, L. 1983. *Southwest Indian drypainting.* Southwest Indian Arts Series. Santa Fe, N.M.: School of American Research and University of New Mexico Press.

Wymer, J. 1982. *The Paleolithic age.* New York: St. Martin's Press.

Wynn, T. 1994. Tools and tool behavior. In *Companion encyclopedia of anthropology,* ed. T. Ingold, 133–61. London: Routledge.

Wynn-Edwards, V. 1962. *Animal dispersion in relation to social behavior.* New York: Hafner.

Zahavi, A. 1975. Mate selection—a selection for a handicap. *Journal of Theoretical Biology* 53:205–14.

Zahavi, A., and A. Zahavi. 1997. *The handicap principle: A missing piece of Darwin's puzzle.* New York: Oxford University Press.

Zervos, C. 1970. *Pablo Picasso, 1969–1970. Exposition concue et mise aupoint par Yvonne Zervos.* Genève: Imprimeries Populaires, Arts Graphiques.

Zimmer, H. 1955. *The art of India and Asia: Its mythology and transformations.* New York: Pantheon.

Ziudema, R. 1973. Kinship and ancestor cult in three Peruvian communities: Hernandez Principe's account of 1622. *Bulletin de l'Institut Français de'Etudes Andines* 2 (1): 15–33.

———. 1992. Inca cosmos in Andean context: From the perspective of the Capac Raymi Camay Quilla feast celebrating the December solstice in Cuzco. In *Andean cosmologies through time,* ed. R. Dover, K. Seibold, and J. McDowell, 17–45. Bloomington: Indiana University Press.

# Index

fathers, 12–13, 14, 79, 102–106
female choice, 107, 118
fictive kin, 6, 16, 123, 140; definition of,
    172. *See also* metaphorical kin
Flo (of Gombe), 12, 13, 80–81, 88, 112
Florence, Italy, 48, 55
foot binding, 161. *See also* body modi-
    fication
Fore, 28
Fortes, Meyer, 12. *See also* axiom of kin-
    ship amity
Fox, Robin, 102
Fra Filippo, 55
Freud, Sigmund, 63
frigate bird, 69, 70

Gadjari, 5. *See also* Australian Aborigines
Galapagos Islands, 145
Gambutis, Marija, 28, 125, 128
genes, 79, 148, 150, 157, 161; good
    genes, 158
genius, 63
genre, 64, 65
Germanic tribes, 53
gild restrictions, 55, 67
Giotto di Bondone, 49, 54
God, 16
Gombrich, Ernst, 5, 53, 73
Goodale, Jane, 99
Goodall, Jane, xii, 80–81, 93, 98, 112
gope boards, 34, *35*
Goya y Lucientes, Francisco, 59, 61
Grandchild Altruism Hypothesis, 91
grandmothers (older women), 14; as art
    teachers, 25–26; as social critics, 100,
    101; as busy bodies; 93; taskmistress,
    93; teachers of kinship behavior, 164;
    the most honored state, 93–94; males
    deferral to, 94
Gravettian, 23
Greece, x, 47, 48, 63, 109; Athens, 49;
    city states, 49; Classical period, 49–
    50; drama, 51; Hellenistic period, 52;
    Homer, 52; Plato's *Republic,* 51, 52;
    Praxitieles, 52; primeval clan organi-
    zation, 48; influence on the Renais-
    sance, 55; influence on Rome, 52;
    Scopas, 52; Sophocles, 51; Theater

of Dionysis, 51; destruction of tradi-
    tions, 49, 51, 52
Grieder, Terence, 125
group benefit selection, 18, 91, 151, 166

hair arrangement, 14. *See also* body
    modification
Hamilton, Annette, 99
Hamilton, William, 149
Hei-tiki, 6, 7, 37
Hiatt, Lester, 99, 101
hierarchies, 14, 15, 20, 136, 168, 169;
    birth order, 93, 136; definition of, 92;
    association with duty, 92, 113; inter-
    ests, 113; maternal, 15, 91–95, 111;
    pecking orders, 92, 102–103, 113
Hindu, 125
Hobbes, Thomas, 91, 92, 102
Hoebel, E. A., 63
Hogarth, William, 59
Homer, 52, 62
Hopi, 23, 93. *See also* American Indians
    of North and South America
Hrdy, Sarah B., 3, 11, 12, 90, 82, 94, 95,
    98, 103–104, 161, 163
huacas, 29, 44, 139, 140, 171
Huston, M. G., 4

idiosyncratic behavior, 11, 124, 160
Inca, 30, 41, 33, 139. *See also* American
    Indians of North and South America
inclusive fitness, 149–150. *See also* kin
    selection
Indonesia, 30
infant care, 30, 86
infanticide, 159
Ingres, Jean Auguste Dominique, 61, 64
inheritance of art motifs, 5
initiation rituals, 142–143, 161
innovation, 16
Inuit, 101. *See also* Indians of North and
    South America
Irish, 94
Ismene, 51

Jamaica, 104–106
Jewish: art, 53, 140; law, 53, 140
Jolly, Alison, 103

Muskogee, 100. *See also* Indians of North and South America

mutation, 146

narrative paintings, 58, 61, 63

natural selection, 145, 152, 157, 166

naturalistic fallacy, 159

Navajo, 24, 94, 120, 165; sand paintings of, 24, 120. *See also* American Indians of North and South America

Neolithic, 28, 127

Neo Platonic, 55

Netsilik Eskimo, 94. *See also* Indians of North and South America

New Caledonia, 41

Nightingale, Florence, xii

noble, 172

Norman Conquest, 54

Northwest Coast, 23. *See also* American Indians of North and South America

nudity, 56

Nuer, 18. *See also* African art

New Caledonia, 41

Nigerian Owo, Benin and Onitsha: 38; *see also* African art

Oedipus, 51

Ona, 23. *See also* Indians of North and South America

oppression of women, 100–101

orgasm, 98

Otten, Charlotte, 72

ovulation (concealed), 103

Pacha Mama, 16

Padju Epat, 28

Palmer, Craig, 111, 118, 121, 159, 162, 163, 166, 171

parental behavior, 13

parental investment, 12

parenting, 79, 81; as k-strategy, 155

Pascua ritual, 9–10

paternal care hypothesis, 103–104

paternity: confidence in, 103

patriarchy: duties to female kin, 102; as life givers, 102; determinant of legitimacy, 102–103; based on superficial evaluation of male role, 100–102

pecking order, 92, 102–103. *See also* hierarchy

penguins, 104

persistence, 109

Petrarch, 49, 54

phenotype, 115, 157, 158

Pima, xi, 25–27. *See also* Indians of North and South America; Salt River Pima/Maricopa Indian Community

Plains Indians, 4, 139–140, 165. *See also* American Indians of North and South America

Plato, 52, 59, 63, 68, 75

Polo, Marco, 33

Pope Gregory the Great, 53

Poro society, 39. *See also* African art

Poussin, Nicolas, 75

Praxiteles, 52

pregnancy, 83–85, 86

propaganda, 52, 109, 117

proximate causes, 3, 147. *See also* cause

Pukeroa Gateway, 41, 42

public health, 117

Quiche, 30. *See also* Indians of North and South America

Quito, Ecuador, 47

r-strategy, 3, 11, 97, 105, 154, 156

Radin, Paul, 2

Rainbow serpent, 4, 16. *See also* Australian Aborigines

rape, 153

Raphael (Raffaello Santi, Raffaello Sanzio), 57

reciprocal altruism, 18, 150–151, 165–166

reciprocity, 172

red ochre, 6, 45, 125. *See also* Australian Aborigines

religion, 21, 120–121

Renaissance, 49, 54–57, 64

reproduction: spontaneous abortion, 84, 85; allomothers limit, 85; cephalo-pelvic disproportion, 84; childbirth, 86; complex process, 83; exterogestate foetus, 88; female, 83–95; lactation as nutrient drain, 87; maternal

# About the Author

Kathryn Coe spent over thirty years preparing to write this book by interviewing contemporary painters, sculptors, photographers, and ceramists, both mainstream and regional. She also worked with indigenous potters, carvers, and weavers in the Southwestern United States, Northern Mexico; small villages in Spain, Colombia's highlands, and Ecuador's paramos, coastal rain forest, and Upper Amazon region. She is assistant professor in the Department of Anthropology at the University of Missouri-Columbia, where she teaches classes in anthropology and art and the health of mothers and families. She is the mother of two children and the grandmother of one.